METAPHYSICS, MATERIALISM,
AND THE EVOLUTION OF MIND

Charles Darwin in 1840. *Courtesy of Nora Barlow.*

Metaphysics, Materialism, & the Evolution of Mind

Early Writings of Charles Darwin

Transcribed and Annotated by
Paul H. Barrett

With a Commentary by
Howard E. Gruber

THE UNIVERSITY OF CHICAGO PRESS

This book originally was published in somewhat
different form as book 2 of *Darwin on Man*.

Acknowledgment is made to the Syndics of Cambridge University
Library for permission to copy Darwin Manuscripts reprinted in this
book. Acknowledgment is also made to the Syndics of the Cambridge
University Press for permission to reprint passages from *Charles
Darwin's Diary of The Voyage of The Beagle*, edited by Nora Barlow
(1934).

The University of Chicago Press, Chicago 60637

© 1974 by Howard E. Gruber and Paul H. Barrett
© 1980 by Paul H. Barrett
All rights reserved. Published 1974
Phoenix Edition 1980
Printed in the United States of America

84 83 82 81 80   5 4 3 2 1

Library of Congress Cataloging in Publication Data

Darwin, Charles Robert, 1809–1882.
   Metaphysics, materialism, and the evolution of mind.

   "This book originally was published in somewhat different form as
book 2 of Darwin on man."
   1. Darwin, Charles Robert, 1809–1882. 2. Evolution. 3. Psychology,
Comparative. 4. Creative ability in science. 5. Naturalists—England—
Biography. I. Barrett, Paul H. II. Gruber, Howard E. III. Title.
QH31.D2A25 1980   155.7   LCN: 80–15763
ISBN 0-226-13659-0 (pbk.)

To Paul H. Jr., Thomas E., and Wilma M. Barrett,
Margaret Barrett Rathert

Contents

ABBREVIATIONS

Frequently used sources will be referred to by the abbreviations listed below. All the manuscripts listed are kept in the Cambridge University Library. In the case of the books, the edition cited below is the one referred to in the text unless otherwise specified.

MANUSCRIPTS

B	Darwin's first notebook on Transmutation of Species (July 1837—February 1838)
C	Darwin's second notebook on Transmutation of Species (February–July 1838)
D	Darwin's third notebook on Transmutation of Species (July 15th 1838—October 2nd 1838)
E	Darwin's fourth notebook on Transmutation of Species (October 1838—July 10th 1839)
M	Darwin's first notebook on Man, Mind, and Materialism (July 15th 1838—October 1st 1838, approximately)
N	Darwin's second notebook on Man, Mind, and Materialism (October 2nd 1838—August 1st 1839, approximately)
"Old and Useless Notes" or [OUN]	"Old and Useless Notes about the moral sense & some metaphysical points" (written mainly from 1837–1840)
Journal	A running record of major events in Darwin's life from 1809–1881. He probably began it in August 1838, the earlier entries being retrospective.

BOOKS

Autobiography	*The Autobiography of Charles Darwin, 1809–1882. With original omissions restored,* edited with appendix and notes by his grand-daughter Nora Barlow (London: Collins, 1958)
LL	*The Life and Letters of Charles Darwin,* edited by his son, Francis Darwin, 3 vols. (London: Murray, 1887)
ML	*More Letters of Charles Darwin,* edited by Francis Darwin, 2 vols. (London: Murray, 1903)

Beagle Diary
Charles Darwin's Diary of the Voyage of H.M.S. "Beagle," edited from the *MS* by Nora Barlow (Cambridge: University Press, 1934)

Voyage, 1839
Journal of Researches into the Geology and Natural History of the Various Countries Visited by H.M.S. Beagle under the Command of Captain FitzRoy, R.N. from 1832 to 1836, by Charles Darwin (London: Colburn, 1839)

Voyage, 1845
Journal of Researches into the Natural History and Geology of the Countries Visited during the Voyage of H.M.S. Beagle Round the World, second edition by Charles Darwin (London: Murray, 1845)

Origin
On the Origin of Species by means of Natural Selection, or the Preservation of Favoured Races in the Struggle for Life, by Charles Darwin (London: Murray, 1859)

Animals and Plants
The Variation of Animals and Plants under Domestication, by Charles Darwin, 2 vols. (London: Murray, 1868)

Descent
The Descent of Man, and Selection in Relation to Sex, by Charles Darwin, second edition (London: Murray, 1882)

Expression
The Expression of the Emotions in Man and Animals, by Charles Darwin, reissue of second edition edited by Francis Darwin in 1890 (London: Murray, 1921)

Zoonomia
Zoonomia; or the Laws of Organic Life, by Erasmus Darwin, 2 vols. (London: Johnson. Vol. I, 1794; Vol. II, 1796). All references are to Vol. I.

SYMBOLS

/ a few words inserted by Darwin /
< crossed out by Darwin >
[added in transcription by Barrett]
[Darwin's own brackets]CD
((marginal or interlinear passage))
| = end of MS page
e = part of MS excised

CHRONOLOGY

Note: Direct quotations are taken from Darwin's *Journal.*

1809 February 12	Born at Shrewsbury.
1818	Entered Shrewsbury School.
1825–27	Attended Edinburgh University.
1827 March 27	Contributed two scientific papers to Plinian Society. W. A. Browne's paper, on mind as material, read and stricken from record at same meeting.
1827–31	Attended Cambridge University.
1829 Summer	Entomological tour of North Wales with Professor F. W. Hope.
1831 Spring	Began planning scientific voyage to Canary Islands.
1831 August	Geological tour of North Wales with Professor Adam Sedgwick.
1831 August 29	Received offer of post of naturalist on H.M.S. *Beagle.*
1831 December 27	H.M.S. *Beagle* sailed from Devonport, England.
1832 September 23	First important fossil find: various extinct mammals.
1832 December 16	First sight of Indians of Tierra del Fuego.
1835 September	Studied geology, fauna, and flora of Galapagos Islands.
1835 December	First draft of paper on theory of formation of coral reefs.
1836 October 2	H.M.S. *Beagle* docked at Falmouth, England.
1837 May 31	Read paper on coral reefs to London Geological Society.
1837 July	"Opened first notebook on 'Transmutation of Species.'" Formulated monad theory of evolution.
1837 October	Began work leading to *Zoology of the Voyage of H.M.S. Beagle, edited and superintended by Charles Darwin,* published 1840–1843, 5 volumes.

1837 November 1   Read paper on earthworms to London
                         Geological Society.

1838 July 15   Began notebooks on man, mind, and ma-
                         terialism.

1838 September 21   Dream of execution.

1838 September 28   Read Malthus, grasped theory of evolution
                         through natural selection.

1838 November–December   Restated theory as three succinct prin-
                         ciples.

1839 January 29   Married Emma Wedgwood.

1839 January–May   Began circulating *Questions About the
                         Breeding of Animals.*

1839 December 27   First child born, William Darwin. Began
                         observations on infant development.

1842, 1844   Wrote preliminary essays, similar in out-
                         line to *Origin of Species.*

1844   Robert Chambers' *Vestiges of the Natural
                         History of Creation* published anony-
                         mously.

1846 October 1   Finished third and last volume of *The
                         Geology of the Voyage of the Beagle.*

1846 October 1   Began eight-year study of barnacles, result-
                         ing in four volumes.

1854 September 9   Finished barnacles.

1854 September 9   "Began sorting notes for Species theory."

1856 May 14   Began writing *Natural Selection,* volumi-
                         nous work on evolution, never finished.

1858 June 18   Received letter from Alfred Russel Wal-
                         lace, formulating theory of evolution
                         through natural selection.

1858 July 1   Papers by Darwin and Wallace, announc-
                         ing theory of evolution through natural
                         selection, read at Linnaean Society, Lon-
                         don.

1858 July 20   Began writing *Origin of Species.*

1859 March 19   Finished writing last chapter of *Origin.*

1859 October 1   Finished correcting proofs.

1859 November 24   *Origin of Species* published.

1860 January 9   "Began looking over MS. for work on
                         Variation."

1860–67   Worked on unsolved problems of variation
                         and heredity, published in 1868 as *The
                         Variation of Animals and Plants under
                         Domestication,* various related botanical

	works on plant reproduction, hybridization, and variation.
1863	Began experimental work on climbing and insectivorous plants, continued until his death.
1867 February	Began circulating questions about expression of emotions. Began work on "Man Essay."
1871 January 15	Finished correcting proofs of *The Descent of Man,* published February 24.
1871 January 17	"Began Expression & finished final rough copy on April 27."
1872 August 22	Finished last proofs of *The Expression of the Emotions in Man and Animals,* published November 26, 1872.
1877	Wrote and published "A Biographical Sketch of an Infant," based on observations made thirty-seven years earlier.
1881	"All early part of year Worm book published Oct. 10th." This was *The Formation of Vegetable Mould Through the Action of Worms, with Observations on Their Habits,* an enterprise begun forty-four years earlier.
1882	Died.

Preface to the Phoenix Edition

Science is a way of life distinct from other activities such as religion, politics, music, literature, and commerce. Yet like all social pursuits, science is inextricably interwoven with the culture as a whole—science cannot exist in a social vacuum—and modern culture would not be the culture it is without science.

The advancement of science and the advancement of human cultures are thus interdependent. No matter how innovative a scientist may be, unless there is a receptive audience, new discoveries are ignored. There are many examples in past history of great scientific breakthroughs going unrecognized by the scientific community or by the public. In some instances the new ideas were at variance with cherished views, and existing power structures could not tolerate threats of major changes in the status quo (Copernicus and Galileo are classic cases); in other instances, contemporary assumptions were incompatible with postulates of the new theories (Gregor Mendel, for example).

Even as a young man, Charles Darwin understood and appreciated functional relationships between the scientific and the nonscientific components of society. His background equipped him to make these crucial judgments; his family life had given him many advantages. He grew up in the presence of a marvelous group of intellectuals. His father, Dr. Robert Darwin, with abundant affection, attention, and warm support, encouraged him to attain an excellent education. Charles learned early in life the qualities necessary to be his own man and, at the same time, to participate easily in the give-and-take of social interaction that leads to psychological maturity.

When we review Darwin's professional accomplishments, and when we put these accomplishments in the context of the often bitter controversies which resulted, we may wonder how Darwin succeeded in conveying the message of evolution to so large and to so hostile an audience when all others who tried had failed. For one thing, Darwin understood both the degree of scientific sophistication and the needs and values of his audience. In order to make these judgments, he drew on the experiences of his own family members as they themselves had struggled with similar problems.

His father, in his daily medical practice, was constantly relating to sick, poorly educated people. He not only diagnosed and treated illnesses, but also dealt with the sick as individuals, each a victim of special circumstances, and each needing compassion, sternness, patience, and often persuasion, to submit to medical attention. At the beginning of his medical career, Robert had had a fierce and prolonged argument with another physician—Dr. William Withering, from nearby Birmingham, best known for his contributions to treatment of heart disease with digitalis. The debate became public, with each doctor vehemently criticizing the other in open letters. From this experience, Robert learned the practical value of professional ethics, namely, that to be "successful," to have many clients and the confidence of those clients, required not only technical competence, but the respect and good will of the community.

Charles's grandfather, Dr. Erasmus Darwin, and uncle (Robert's brother, also named Charles Darwin) had both been in the medical profession and had both received considerable recognition for their achievements in science. Erasmus had also become famous as a physician and a poet. Although Charles's uncle had his life cut short at the age of twenty, due to an infection brought on by a tragic accident during an autopsy, he had been a promising young scientist and, even while a student at Edinburgh University, had earned high awards, impressing his professors with his talent and good judgment. Erasmus, the grandfather, was widely known throughout Britain for his views on evolution, and had it not been for his reputation as a physician and philosopher, might easily have been persecuted as was his friend and fellow Lunar Society member, Joseph Priestley. To believe in the evolution of species was considered heretical, and to support the theory in print, as Erasmus did, was to invite severe condemnation and violent attack.

In early 1837, Charles Darwin began work on his theory of evolution. He was twenty-eight years old; he knew that the route ahead was long, tortuous, lonely, and hazardous. But his mind was set, and there was no thought of retreat. He was well prepared for the task; he had just returned from a five-year trip around the world as naturalist on board H.M.S. *Beagle*. His experiences in South America and in Australia, and on hundreds of islands in the main oceans of the world, had convinced him that evolutionary laws explained the geographical distribution of species, their fossil forms, their family relations, and numerous other facts better than did views of creation based on ancient biblical writings.

It is unlikely that anyone could have had better qualifications than Charles to meet the awesome challenge of proving this theory. As a youngster, from his tenth to sixteenth year, he had attended Shrewsbury School. This was the best boys' school in all of Britain, judged by the number of its graduates earning scholarships and awards at Cambridge and Oxford Universities. The headmaster, Dr. Samuel Butler, was himself a dedicated and productive scholar, and he developed a school with high standards. His pupils were well schooled in the classics, including Latin and Greek, and in literature, geography, history, and arithmetic. At the age of sixteen Charles entered the University of Edinburgh, Scotland, where he studied medicine for two years. He then studied for the ministry at Cambridge University, where in 1831, after three years, he earned his bachelor's degree. At Shrewsbury, at Edinburgh, and at Cambridge he had studied the humanities and the sciences under the most outstanding educators to be found anywhere. He had come from a home where books were common and their use encouraged, and where intellectual discussion thrived. He was at ease with university dons and with political and commercial leaders, but he was also readily accepted and respected by ordinary people.

No later than the month March 1837, Darwin became convinced that evolution was true. He had been writing his ideas about this and other subjects in a small notebook. On the front of the notebook he wrote, in large letters, "RN." His jottings were about geology, and about different species, and included speculations that species change from one kind to another. That spring, apparently in July, he started two other notebooks, labeled "A" and "B," in which he restricted his notes to geology and to species respectively. By February of 1838 the "B" notebook was filled, and he continued the species notations in a new notebook labeled "C." The second species notebook was filled sometime in July, and on July 15, 1838 he started a third notebook, "D," on the subject.

It is important to realize 'that in keeping these notebooks Darwin was carrying on a procedure that has always been considered valuable if one is to produce significant and reliable scientific results. It was probably not difficult for Darwin to discipline himself to write down his thoughts in a systematic and orderly manner, for he had been accustomed from early in life to writing letters to family members and friends and had kept diaries, when a student at Edinburgh, recording observations on natural history. He had filled dozens of notebooks during the voyage of the *Beagle*.

The notes that Darwin wrote in his species notebooks are of various kinds. For one thing, they represent his private thoughts, since he had no reason to fear exposure of their contents. In most cases the notes are just that. He did not attempt to compose finished drafts of text material; the words he recorded are jottings, disconnected thoughts, single phrases, clues to fleeting thoughts. He recorded ideas, however unpromising, whether they were his own or someone else's. He listed bibliographic sources as he encountered them in his readings; he noted fragments of conversations with his father, colleagues, friends, and strangers.

In July of 1838, as he opened his third notebook on species, Darwin apparently turned a major corner in his approach to the evolution problem. He decided to separate his notes on the subject and simultaneously started another notebook. On the front of this other notebook he wrote "M." Nearly twenty years later, looking over the M notebook, he wrote on the inside front cover, "This Book full of Metaphysics on Morals & Speculations on Expression." By the end of September 1838, he had filled the D notebook on species and the M notebook on metaphysics.

On September 28, 1838, in the D notebook on pages 134 and 135, Darwin wrote his famous notes about Malthus (see pp. 195–196 in this book).[1] In early October he started, simultaneously, two new notebooks, "E" and "N." Years later, on the inside of the N notebook, he wrote, "Metaphysics & Expression for Species Theory." The last dated entry in the E notebook is July 10, 1839; in the N Notebook there are entries dated July 20, 1839, and April 28, 1840.

Today, the B,C,D, and E series is known as Darwin's *Transmutation Notebooks*, and the M and N series as the *Metaphysics Notebooks*. During the period 1837 to 1842, Darwin recorded evolutionary and geological notes in at least one other notebook, and on miscellaneous scraps of paper. One group of papers was put into an envelope and labeled "Old and Useless Notes about the Moral Sense & Some Metaphysical Points." Another notebook on species was later entirely cut up, and today is called the "Torn-up Notebook." Sometime in the fall of 1838, soon after he read Malthus, he read a book entitled, *Proofs and Illustrations of the Attributes of God*, by John Macculloch, and wrote a draft of his impressions of the book. A transcription of this essay is published here under the title "Essay on Theology and Natural Selection." Transcriptions of the M, N, and "Old and Useless Notes" (OUN) are also published here. Finally, since the principal subjects of M, N, OUN, and "Essay on Theology and Natural Selection" are man, mind, and materialism, extracts on these subjects have been included in this book from Darwin's *Beagle Diary* and from the *Transmutation Notebooks*. Finally, an article published in 1877 entitled "A Biographical Sketch of an Infant" is reprinted here. Darwin wrote this article thirty-seven years after having made notes on the growth and development of his first child, William, who was born in 1839. Many observations and ideas on the expressions of emotions, and on the evolutionary origin of these phenomena, are presented in this article.

Darwin never said, so far as we know, why he separated his evolutionary notebooks into the transmutation and metaphysics series. It seems clear, however, that he was collecting notes even at that early date with the purpose of eventually publishing his theories. He knew that any theory of evolution would be attacked, and that discretion demanded that he keep references to human evolution to a minimum. Apparently, even in these early stages, he began two books, one on a general theory of evolution, and a second on the evolution of man and mental faculties. The latter work would be held in abeyance, pending the success of the first.

His well-planned preparations bore abundant fruit, as history has proved. The transmutation notebooks went through four stages of enlargement and revision—becoming, in 1859, *The Origin of Species*. The first revision was the "Sketch of the Origin of Species," written in 1842; the second was the "Essay for the Origin of Species," written in 1844.[2] The third was a manuscript for the "Big Book on the Origin of Species by Natural Selection," written sometime between 1856–1858.[3]

The metaphysics notebooks, along with OUN and ancillary materials, were expanded in later years into chapter 7, "Instinct," in *The Origin of Species* (1859); *The Descent of Man, and Selection in Relation to Sex* (1871); and *The Expressions of Emotions in Man and Animals* (1872). Also, numerous books and articles were based on the M and N notebooks: "The Movements and Habits of Climbing Plants" (1856); *Insectivorus Plants* (1875); *The Power of Movement in Plants* (1880); *The Formation of Vegetable Mould, through the Actions of Worms, and Observations on their Habits* (1881); "Inherited Instincts" (1873); "Habits of Ants" (1875); and "Movements of Leaves" (1881).[4]

Professor S. S. Schweber of Brandeis University has said, "To the best of my knowledge the M and N notebooks contain the first presentation of an evolutionary view of society based on an evolutionary view of nature." He also believes that "the M and N notebooks and the Old and Useless Notes are . . . an account of Darwin's search for God."[5]

Professor Howard Gruber of Rutgers University has discussed the historical background leading to Darwin's evolutionary studies.[6] He has, in addition, provided a running commentary for Darwin's notes in this book. The commentary is a useful interpretation and analysis of the fragmentary and often obscure jottings, particulary for the reader not well aquainted with Darwin's style.

In order to explore the subjects of materialism, metaphysics, and the evolution of the mind, Darwin had to keep his thoughts and notes secret. To do otherwise especially at a time so early in his career, would have jeopardized his scientific reputation and virtually guaranteed his professional demise. Consider the impact the following statements would inevitably have had, were they to have been made public: "I verily believe free will & chance are synonymous.—Shake ten thousand grains of sand together & one will be uppermost,—so in thoughts, one will rise according to law." (*M* 31, p. 11) "With respect to free will, seeing a puppy playing cannot doubt that they have free will, if so all animals, then an oyster has . . . If so free will is to the mind, what chance is to matter." (*M* 72, p. 18) "The above views would make a man a pre-destinarian of a new kind, because he would tend to be an atheist." (*M* 74, p. 19)

As Darwin collected his notes, many lines of inquiry opened. He scribbled his thoughts down as fast as possible—and the result, even today, is almost overwhelming. He had set for himself a nearly impossible goal—to try to dismantle a traditional, normative, deeply entrenched world view, and to supply a substitute. If the human species had come about as the result of the evolution of biological organisms through eons of time, changing, adapting, all under the control of physical laws of the universe, then why invoke spiritual forces?

The challenge was to show how animals, nonhuman animals, and indeed plants were part of a single, ongoing evolutionary process. What particular questions arose? What speculations tied the various pieces of the big puzzle together? He was already satisfied that humans, or at least the human body, had evolved from prehuman progenitors. But had the mind and those so-called unique human attributes—such as morality, intelligence, creativity, social conscience, free will, deliberation, the sense of guilt, artistic taste, cooperation, and all those other special mental faculties of which humans boast—evolved similarly?

In coping with these large, all-encompassing questions, Darwin moved a multitude of smaller ones into focus. He was not only years ahead of his time, but, in many ways, he was ahead of our time, for we still haven't solved many of the mysteries he articulated. Here is a list of questions he dealt with in the notebooks: Are certain hereditary traits linked? Are thought processes brain secretions? Is brain structure altered by thought action? Do mental illnesses have adaptive value (do they contribute to survival)? What value judgments are appropriate in distinguishing normal from abnormal behavior? What is the relative value of animistic versus materialistic explanations of mind action? What are the physiological and psychological functions of sleep, dreams, and senility? Are congenital anomolies hereditary? To what extent can free will explain behavior? Are creative, cognitive, and other mental faculties hereditary? Is the esthetic sense hereditary? Do science and the arts have in common the sense of beauty, poetry, and rhythm? Is forgetfulness therapeutic? Do habitual actions become hereditary? Why are emotional opposites often paired, for example, laughing/crying, pain/pleasure? Does instinctive behavior vary inversely with intelligence? Is human language derived from communicative symbols of progenitors? Do simple animals (oysters, earthworms) reason? Does the moral sense vary in different races of mankind? To what extent are emotions under the control of will power? Is happiness of adaptive advantage in the struggle for survival? Is the notion of property inherited? Are artistic traits inherited? Is the sense of morality a biological form of social adaptation which is derived by evolutionary processes?

As the notebooks illustrate, Darwin set about the task of researching the issues of metaphysics, materialism, and mind with all the vigor, intensity, and thoroughness typical of everything he tackled. The notebooks are an invaluable source of information about the scope of his investigations, and they are also an excellent example of a scientist at work. We see the power of scientific methodology in its mature form: the collection of empirical data, generalization by induction, construction of a new world view, deduction of particular facts new to man's thinking, formation of new theories by imagination and speculation, prediction, and suggestions for tests and experiments. Darwin was no slave to the early nineteenth-century dictum that science was strictly an exercise in observation.

At one point Darwin made a note to himself to go through a dictionary and "make list of every word expressing a mental quality." (*N* 88, p. 86) We have no evidence that Darwin ever actually compiled such a list; however, by means of the modern computer, it is easy today to collect the words that he used to express emotions and other mental qualities. A concordance for the notebooks produced at Michigan State University has made it possible to examine this part of Darwin's vocabulary. A few examples from the list of 229 words gives the flavor of the index, and indicates Darwin's attention to detail: abhorrence, admire, affection, afraid, alacrity, amusing, anger, anguish, ashamed, avarice, bashful, betray, chaste, delusion, depressed, despair, despise, desultory, dream, envy, fickle, fright, grief, happy, hostile, imitate, insane, joy, love, mania, moral, pain, peevish, pity, pleasure, pout, rage, sad, sagacity, satirical, satisfied, sceptical, sleep, snarl, sneer, vicious, weariness, wearisome, whine, wonder, and yawn.

Darwin was no less thorough in studying a wide variety of nonhuman species than he was in investigating humans. If one were to demonstrate the unity of all sentient beings, then facts must be collected showing the similarities (as well as the differences) among these individuals. Once again a concordance is useful; 136 references to different species occur in 89 pages of Darwin's notes. The list contains, as a representative group, the following: acanthosoma, anastatica, ants, apple, ass, baboon, bats, beetle, *Birgos*, bulls, butterfly, cat, chameleon, chicken, chimpanzee, coral, cow, cowslip, crocodile, dahlia, dingo, dog, earwig, elephant, eucalyptus, fleas, *Fucus*, goose, hawk, hollyhock, hyaena, hydra, iris, jackal, kangaroo, monkey, oyster, peacock, peccary, pig, porpoise, puma, salmon, sheep, spider, stallion, terebrantia, tiger, turkey, verbena, wasp, woodpecker, worms, zebra, zoophyte.

In recent years, due in large part to the work of Professor Edward O. Wilson of Harvard University, sociobiology has received considerable attention. Proponents of certain views in this science suggest that human social behavior is largely under genetic control and results from evolutionary development. Darwin was a pioneer in these studies, as the notebooks show. Some of his views were later construed to be what is now labeled "Social Darwinism" and are still the subject of intense debate. At issue are whether the different races of man vary in their competitive superiority; whether different races possess different mental faculties; whether mental faculties, such as moral value systems, are manifestations of biological survival mechanisms; whether aggressive tendencies are instinctive; and whether aggressive tendencies are universal.

The fact that Charles Darwin could and did confront these potentially explosive subjects, if only in the privacy of his thoughts and notebooks, cannot but arouse in us a sense of awe and admiration. His was clearly an unfettered mind, a free spirit, perhaps even a dramatic contradiction of the belief that the human mind is, as Darwin himself mused, a mechanistic instrument programmed to function within normative constraints of hereditary and cultural restrictions.

The list of noteworthy figures from literature, art, science, medicine, philosophy, political economy, and industry whose contributions Darwin considered worthy of serious study, and whose names he mentioned in the metaphysical notes, reads like a "Who's Who" of the late eighteenth and early nineteenth centuries: Archibald Alison, John Audubon, Francis Bacon, David Brewster, Edmund Burke, Samuel Taylor Coleridge, Auguste Comte, William Cowper, Georges Cuvier, Erasmus Darwin, Humphry Davy, Charles Dickens, Oliver Goldsmith, David Hartley, John Henslow, John F. W. Herschel, Alexander Humboldt, David Hume, John Hunter, J. B. Lamarck, Gotthold Lessing, John Locke, Charles Lyell, James Mackintosh, Thomas Malthus, Harriet Martineau, James Monboddo, A. A. Necker, William Paley, A. Quetelet, Thomas Reid, Johsua Reynolds, Walter Scott, B. Smart, Adam Smith, Dugald Stewart, John Horne Tooke, William Whewell, and William Wordsworth.

To say that Darwin did not do his homework, and to say that his academic education had been "wasted," as he himself put it, would be to deny the overpowering evidence to the contrary.

There is no need to present here an extended analysis of the content and significance of Darwin's metaphysical notebooks. Several authors have already done so. Gruber's contributions have been mentioned. Others whose works are recommended are Silvan Schweber, Edward Manier, Sandra Herbert, David Kohn, Michael Ruse, Sydney Smith, C. U. M. Smith, and Charles Swisher.

The Darwin notes on metaphysics, materialism, and the evolution of mind give us a rare opportunity to examine the workings of one of mankind's greatest intellects. The examination of these notes becomes a study of creativity. Unless he resolved the conflict between materialistic and normative perceptions of the living world, Darwin didn't see how his conclusions about biological evolution could seem reasonable. The need to face these philosophical issues before proceeding further with transmutation investigations was a reflection of Charles Darwin's integrity, his sense of duty to thoroughness and truthfulness. The set of notebooks is a two-year, day-by-day record of his efforts to construct a theory that was to shake the intellectual world. By 1840 Darwin abandoned further serious examination of the metaphysical aspects of evolution. He had become convinced that spiritual forces had no role in evolution.

A word of caution about attempting to date Darwin's notes is in order. It is generally safe to assume that textual notes within a particular notebook were written within the period as dated by Darwin himself. However, the dating of annotations later added by Darwin in the margins of the pages, or between lines, are difficult, if not impossible, to ascertain. In order to decipher the periods in which Darwin made these annotations, expert analyses are needed of, for example, the chemical composition of the various grades of ink and pencil which Darwin used at different times—data difficult, if not impossible, to obtain. Dr. Sydney Smith of Cambridge University, however, is making progress on this sort of analytical work, basing his dating codes on ink colors, pencil traits, etc.

The Appendix of this book includes two small papers of importance to Darwinian history: "A Biographical Sketch of Charles Darwin's Father," and "Plinian Society Minutes Book: Meeting of 27 March 1827, Extracts from page 57." The latter is a transcription of a paper presented by a fellow medical student of Darwin's and is of special interest. The paper was expunged from the minutes because of its "materialistic" views. See Gruber's book, *A Psychological Study of Scientific Creativity*, for a discussion of this document.

Another small item is "Questions for Mr. Wynne." Although Mr. Wynne has not been positively identified, he was probably Sir Watkin Williams Wynn of Wynnstay, North Wales (not far from Chirk Castle, the home of the Myddeltons, and near Woodhouse, the home of the Mostyn-Owens, both families well known to the Darwins). I am indebted to Miss Sara PughJones, of Llangollen, for identification of Mr. Wynne.

NOTES TO PREFACE

1. See also "Darwin's Notebooks on Transmutation of Species, Part VI: Pages Excised by Darwin," edited by Gavin de Beer, M. J. Rowlands, and B. M. Skramovsky, *Bulletin of the British Museum (Natural History)*, Historical Series, Vol. 3, 1967, pp. 129–176. Particular acknowledgment for their contributions to the entire series of "Darwin's Notebooks on Transmutation of Species" is due de Beer, Rowlands, and Skramovsky, See *Bulletin of the British Museum (Natural History)*, Historical Series, Vol. 2, 1960–61; Vol. 3, 1967. In addition see "A Transcription of Darwin's First

Notebook on 'Transmutation of Species,' " edited by Paul H. Barrett, *Bulletin of the Museum of Comparative Zoology* at Harvard University, Vol. 122, 1960, pp. 247–296.

2. Charles Darwin, *The Foundations of the Origin of Species: Two Essays written in 1842 and 1844*, edited by Francis Darwin (Cambridge: Cambridge University Press, 1909).

3. R. C. Stauffer, *Charles Darwin's Natural Selection: Being the Second Part of His Big Species Book Written from 1856 to 1858* (Cambridge: Cambridge University Press, 1975).

4. For complete bibliographic citations see R. B. Freeman, *The Works of Charles Darwin: An Annotated Bibliographical Handlist* (Hamden: Archon, 1977), and Paul H. Barrett, *The Collected Papers of Charles Darwin* (2 volumes, Chicago: University of Chicago Press, 1977).

5. Sylvan S. Schweber, "The Origin of the *Origin* Revisited," *Journal of the History of Biology*, Vol. 10, 1977, pp. 232 and 233.

6. Howard E. Gruber and Paul H. Barrett, *Darwin on Man: A Psychological Study of Scientific Creativity* (New York: Dutton, 1974; Chicago: University of Chicago Press, forthcoming).

PART I

Previously Unpublished Manuscripts of Charles Darwin

The Notebooks on Man, Mind and Materialism

> Nothing can contribute more to obviate the incon-
> venience and difficulties attending a vacant or wan-
> dering mind, than the arrangement and regular dis-
> posal of our thoughts in a well ordered and copious
> common-place book . . .

> John Locke, *Letters on Study* as quoted in the
> printed introduction to the common-place book
> used by Erasmus Darwin: *Bell's Common-Place
> Book, Formed Generally upon the Principles
> Recommended and Practiced by Mr. Locke*
> (London: John Bell, 1770) , p. 5

When we say that someone has "discovered" something, we rarely mean that he is the first human being to have encountered or experienced it. Rather, we mean that the discoverers are the first to become aware of the significance of the thing in question, and the first ones able to communicate their understanding to some larger human group. Only in some such sense can we say that Columbus or any other European explorer "discovered America." Only such an enlarged concept of the process of discovery can explain why important things like continents, great theories, and the physical elements themselves must be discovered over and over before they take their places as stable components of an image of the world shared by a human group. The significance of that which is discovered is always complex and many-layered and multi-faceted; the process of discovery is correspondingly complex, extended over time, and the collective result of several efforts in which under-standing is successfully deepened. Was America discovered by Euro-peans when it became known to sea-farers that if one sailed far enough in a certain direction one encountered land, or when it was recognized that the land in question was much larger than an ordinary island, or when it was realized that the land mass was *not* the "Indies," or when it was understood that beyond it lay a great ocean, etc.?

Cynopithecus niger, in a placid condition. Drawn from life by Mr. Wolf.
The same, when pleased by being caressed. Illustrations from *The Expression
of the Emotions in Man and Animals.*

In a narrow sense, of course, neither we nor anyone else discovered
the M and N notebooks. Darwin wrote them and his family knew
about them. By 1957 they were well enough catalogued in the Library
at the Cambridge University for one of us to have acquired a microfilm
of the manuscripts. By 1960 that catalogue was printed.* Others interested in Darwin and the history of science have certainly held these
notebooks in hand and struggled with Darwin's handwriting and with
his ideas.†

In a larger sense, we are still only mid-course in the process of discovering these documents. Their significance for Darwin and for the
history of ideas is so intricate and profound that no single treatment
can hope to penetrate all the crannies of Darwin's thought or form an
adequate image of the structure of ideas that gave rise to these notes.

The M and N notebooks are fragmentary, disjointed running
records of Darwin's thoughts in certain domains, those connected with

---

* *Handlist of Darwin Papers at the University Library, Cambridge,* published at
the Cambridge University Press, 1960. The M and N notebooks are described in
five lines.

† Those known to us personally who have examined the M and N notebooks include Nora Barlow, Harold Fruchtbaum, Sydney Smith, Stephen Toulmin, and
June Goodfield. Charles N. Swisher has given an account of their contents in his
paper "Charles Darwin on the Origins of Behavior," *Bulletin of the History of
Medicine,* Vol. 41, 1967, pp. 24-42.

man, mind, and materialism. The "transmutation notebooks," out of which the M and N notebooks grew, were begun in July 1837. By July 1838 Darwin felt the need to collect his thoughts on the subjects of man, mind, and materialism—topics which had been increasingly intruding themselves into the transmutation notebooks.

M Notebook

Charles Darwin, Esq.
36 Grt. Marlborough St^r.

PRIVATE

Finished October 2d.

This Book full of Metaphysics on Morals
and Speculations on Expression—
1838
Selected Dec. 16, 1856

(p. 64, on <sweet> ants getting on Table. Col. Sykes.) 1

1   July 15, 1838. My father says he thinks bodily complaints/& mental disposition/oftener go with colour, than with form of body.—thus the late Colonel Leighton[2] resembled his father in body, but his mother in bodily & mental disposition.—

My father has seen innumerable cases of people taking after their parents, when the latter died so long before, that it is extremely improbable that they should have imitated. When attending Mr. Dryden Corbet,[3] he could not help thinking he was prescribing to his father & old Mrs. Harrison, said, although constantly seeing

2   him, she was often struck with this fact.—The resemblance | was in odd twiching of muscles, & general manner of holding hands, etc., etc.—Mr. Dryden Co said he could not remember his father.—

My father thinks people of weak minds, below par in intellect frequently have very bad memories for things which happened in early infancy[4]—of this fact Mr. Dryden C. is good instance as he is very deficient,—he was nearly 9 years old when his father died,—

The omnipotence of habit is shown about meals, no |

3–6e [excised, not found]

7   There is a case of Mr. Anson who told a story of·hunting /— habitual fits—/ which my Father thinks is mentioned in the Zoonomia.—[5]

Now if memory /of a tune & words/ can thus lie dormant, during a whole life time, quite unconsciously of it, surely memory from one generation to another also without consciousness, as instincts are, is not so very wonderful.

<Now is not epilepsy an *habitual* disease of the muscles???> |

8   Miss Cogan's memory of the tune, might be compared to birds singing, or some instinctive sounds.—Miss C. memory cannot be called memory, because she did not remembered. it was an habitual action of thought-secreting organs. brought into play by morbid action.—Old Elspeth's /in Antiquary/ power of repeating poetry in her dotage is fact of same sort.[6] Aunt B.[7] ditto.— |

9   Case of Mr. Corbet of the <Hall> Park,[8] after paralytic stroke. intellect impaired. Could converse well on any subject when once started,—could receive a new train[9] through eyesight, though not through hearing,—Thus when dinner was announced he could not understand it, but the watch was shown him.— ((the servant showed him watch & said dinner is ready, what, what,—then showed the watch upon which he exclaimed, why it is dinner time.—)) My father asked him whether he had gardener of name A. B., etc., etc. & he maintained he had never heard of such a man & had no gardener.—My F. then asked Mr. C. to come to the window & pointed out the Gardener & said, who is that? Mr. C.

10   answered | why do you not know, that is A. B., my gardener.— Thus was he in every respect, no communication could be held by means of hearing.—

Mr. Corbett, however, in conversation could catch up a new train if *early* association were called up.—My F. asked him did he know whom Mr. Child /of Kinlett/[10] had married.—Answered never heard of such a man.— (My Father explained who he was & all about him, but still maintained he had never heard of him.)—

11   My F. then said you remember | Jack Baldwin at school.—Answered To be sure I do.—What became of him.—Answ Had large fortune left him, took name of Child /of Kinlet/ & married Miss A. B.—all the same names as a few minutes before he maintained he had never heard of.—Thus in many things if he began at one end, he knew the whole subject.—if at the other nothing.—He could repeat the alphabet straight, but did not know [Z][CD] when heard isolately.[11] |

12   In old people (Aunt B.) when they hear a thing it often does not take any effect at the time, but some time afterwards it calls up

pain or pleasure & is often recurred to & mentioned as a thing which had just taken place.—As if the idea of time had been disturbed.—

These foregoing cases of /mental/ failure very general effect of <early> /slight habitual/ intemperance.—often accompanied by extreme anger, at not being understood.— |

13   My F. says there is perfect gradation between sound people and insane.—that everybody is insane at some time. Mania is quite distinct, different also from delirium, a peculiar complaint stomach not acted upon by Emetics.—people recognized,—sudden changes of disposition, like people in violent intoxication, often ends in insanity or delirium.—In Mania all idea of decency & affection are lost.—Most delicate people do most indelicate actions, —as if /their emotions/ acquired.—This may be doubted, whether rather not going against natural instincts.—

14   My Grand F. thought the feeling of anger, which rises almost | involuntarily when a person is *tired* is akin to insanity. ((I know the feeling also of depression, & both these give strength & comfort to the body.)) I know the feeling, thinking over somebody who has, perhaps, slightly injured me, [illegible] speeches, yet with a sort of consciousness not just. From habit the feeling of anger must be directed against somebody.—Have insane people any misgivings of the unjustness of their hatreds, as <if> in my case.— It must be so from the curious story of the Birmingham Doctor,[12] praising his sister who confined him. & yet disinheriting her.—

15   This | ((N.B. I have read paper somewhere on horse being insane at the sight of anything scarlet.—dogs ideotic.—dotage.—)) Doctor communicated to my grandfather his feeling of consciousness of insanity coming on.—his struggles against it, his knowledge of the untruth of the idea, namely his poverty.—his manner of curing it by keeping the sum-total of his accounts in his pocket, & studying mathematics.—My Father says after insanity is over people often think no more about it than of a dream.— |

16   Insanity is produced by moral causes (ideally by fear. Chile earthquakes) in people, who, probably otherwise would not have been so.—In Mr. Hardinge was caused by thinking over the misery of an illness at Rome, when by *accidental* delay of money, he was /only/ NEARLY thrown into a hospital.—My father was nearly drowned at High Ercall,[13] the thoughts of it, for some years after,

17 was far more painful than the thing itself. | Asked my F. whether insanity is not distinguished from whims passion etc by coming on suddenly. Ans. no,—because often, if not generally, does not really come on suddenly.—Case of Mrs. C. O. who threw herself out of the window to kill herself from jealousy of husband connection

with housemaid two years before, to prove she was not insane, answered she had known it at time & had bought arsenic for that purpose.—This found to be true.—Her husband never suspected during these two years that she had been insane all the time.— |

18   ((Case of Shrewsbury gentleman, unnatural union with turkey cock, was *restrained* by remonstrances on him.)) There are numberless people insane of particular ideas, which are never generally, if at all discovered.—Sometimes comes on suddenly from (in one case ipecacuhan[14] not acting), in others from drinking cold drink. —then brain affected like getting suddenly into passion.—There seems no distinction between enthusiasm passion & madness.—ira furor brevis est.[15] My father quite believes my grand F. doctrine is

19 true, that the only cure for madness is forgetfulness.— | which does appear a real difference, between oddity & madness.—but then people do not well recollect what they have done in passion.—

People are constantly well aware that they are insane & that their idea is wrong.— (Dr. Ashe, the Birmingham Doctor), in this precisely like the passion, ill-humour & depression which comes on from bodily causes.—

It is an argument for materialism, that cold water brings on suddenly in head, a frame of mind, analogous to those feelings, which may be considered as truly spiritual.— |

20 A person twitching when a disagreeable thought occurs, is closely analogous to Epilepsy & convulsion.—affections of the thinking organs.—the action of brain which gives sensation of pain, emits its power on the muscles in the twitching.

Pride & suspicion are qualities, which my F says are almost constantly present in people, likely to become insane.—now this is well worth considering, if pride & suspicion can be well understood. |

21 In insanity, the ideas do not go back to childhood, (but appear most capricious) as in delirium after epilepsy, but in the failing from old age, they constantly do.—In Mrs. P. of B. thought herself near Drayton & Ternhill,[16] (where she was born) though she never naturally talked of these places.—My F. says shows that early impressions are most durable.—) but Miss Cogan shows that repetition is not necessary) —the words second childhood full of meaning:—Dreams do not go back to childhood—People, my

22 Father says, do not dream of what they | think of *most* intently.— criminals before execution.—Widows not of their husbands—My father's test of sincerity.—

People in old age exceedingly sharp in some things, though so confused in others.—Mrs. P. when in state as above described, (forgetting that her husband was dead) yet instantly perceived when my Father to distract her attention took her /left/ hand to pretend to feel her pulse.—

What fails first?—How is this? Does memory bring in old ideas |

23   Dogs take pleasure, when doing what they consider their duty,— as carrying a basket, bringing back game, or picking up a stone, though only acquired rules by art.—like the law of honour.—they feel pleasure in obeying their instincts naturally.— (generosity in defending a friendly dog).—they feel shame, when doing anything which is wrong.—as eating meat, doing their dirt, running

24   home.—in these cases their actions do not look like | fear, but shame.—I cannot remember instances, but I feel sure I have seen a dog doing what he ought not to do, & looking ashamed of himself.—Squib[17] at Maer, used to betray himself by looking ashamed before it was known he had been on the table,—guilty conscience. —Not probable in Squib's case any direct fear.—

My father thinks that selfishness, pride & kind of folly like (Mr. George S.) is very hereditary.— |

25   My Father says on authority of Mr. Wynne,[18] that bitch's offspring is affected by previous marriages with impure breed.—

A cat had its tail cut off at Shrewsbury & its kittens (in number 3) had all short tails: but one a little longer than rest /they all died/:—she had kittens before & afterwards with tails.

My father says, perfect deformity, as an extra number of fingers,—harelip or imperfect roof to the mouth (as in Lord Berwick's[19] family) ((stammering in my Father's family))[20] are hereditary.—

Other deformities are illnesses of the foetus.—some mothers, have first dead children, then children which were short term, & lastly healthy ones.— |

26   Insanity & Epilepsy remain many generations in families.—My father does not know whether trains of insanity are hereditary in any one family.—

In Aunt B. the affections ((N.B. affections very soon go in Maniacs)) seem to have failed even more than the memory.— therefore affections effect of organization which can hardly be doubted when seeing Nina with her puppy.—The common remark that fat men are goodnatured, & vice versa Walter Scotts remark how odious an illtempered fat man looks, shows same con-

27   nection between organization & | mind.—thinking over these things, one doubts existence of free will every action determined by hereditary constitution, example of others or teaching of others.— (NB man much more affected by other fellow-animals, than any other animal & probably the only one affected by various knowledge which is not hereditary & instinctive) & the others learnt

what they teach by the same means & therefore properly no free will.—we may easily fancy there is, as we fancy there is such a thing as chance.—chance governs the descent of a farthing, free will determines our throwing it up,—equally true the two statements.— |

28   Catherine[21] remarks that pleasure received from works of imagination very different from the inventive power; this, though very odd is perhaps true.—Mem Erasmus & mine taste for music.—Children like hearing a story told though they remember it so well that they can correct every detail, yet they have not imagination enough to recall up the image in their own mind.—this may be worth thinking over.—it will perhaps show differences between memory & imagination. ((Catherine thinks that children like looking at pictures, an easy task, of animals they know.—pleasure of imitation (common to monkey) , & not imagination.—)) |

29   Thinking over the scenes which I *first recollect* /at Zoos/, they are all things which are brought to mind, by memory of the scenes, (indeed my American recollections are a collection of pictures) .— when one remembers a thing in a book, one remembers the part of page.—one is tempted to think all memory consists in a set of sketches, some real—some fancied.—this fact of early memory consisting of things seen, quite agrees with my Fathers case of Mr. Corbet of the Hall understanding, (on hearing old association brought up) by sight & not by hearing |

30   One is tempted to believe phrenologists are right about habitual exercise of the mind, altering form of head, & thus these qualities become hereditary.—When a man says I will improve my powers of imagination, & does so,—is not this free will,—he improves the faculty according to usual method, but what urges him,—absolute free will, motive may be anything ambition, avarice, etc., etc. An animal improves because its appetites urge it to certain actions,

31   which are modified by circumstances, & thus the | appetites themselves become changed.—appetites urge the man, but indefinitely, he chooses (but what makes him fix!? frame of mind, though perhaps he chooses wrongly,—& what is frame of mind owing to—) I verily believe free will & chance are synonymous.—Shake ten thousand grains of sand together & one will be uppermost,—so in thoughts, one will rise according to law.[22]

How strange <all> /so many/ birds singing in England, in Tierra del Fuego not one.—now as we know birds learn from each other /though different species/ when in confinement, so may

32   they | learn in a state of nature.—Singing of birds, not being instinctive, is hereditary knowledge like that of man, & this agrees with the stated fact, that /birds from/ certain districts have the

best song. [Migratory birds return to same quarter for many years].^{CD}—

Beauty is instinctive feeling, & thus cuts the Knot:—Sir J. Reynolds[23] explanation may perhaps account for our acquiring /the *instinct*/ one notion of beauty & negroes another; but it does not explain the *feeling* in any one man.— |

33   Music & poetry opposite ends of one scale.—former pleases from instinct the ears (rhythm & *pleasant sound per se*) & causes the mind to create short vivid flashes of images & thoughts.—Poetry. the latter thoughts are in same manner vivid & grand, the frame of mind being just kept up by the music of the poetry.— (therefore singing intermediate. Who has not had his blood run cold by singing) .—

Granny[24] says she never builds castles in the air—Catherine often, but not of an inventive class.— |

34   Now that I have a test of hardness of thought, from weakness of my stomach I observe a long castle in the air, is as hard work (abstracting it being done in open air, with exercise etc, <not> no organs of sense being required) as the closest train of geological thought.—The capability of such trains of thought makes a discoverer, & therefore (independent of improving powers of invention) such castles in the air are highly advantageous, before

35   real train of inventive | thoughts are brought into play, & then perhaps the sooner castles in the air are banished the better.— The facility with which a castle in the air is interrupted & utterly forgotten, so as to feel a severe disappointment because train cannot be discovered ((in real train of thought this does not happen. because papers, etc., etc. round one. one recalls the castle by going to the beginning of castle)) —is closely analogous to my Father's positive statement that insanity is only cured by forgetfulness.— & the approach to believing a vivid castle in the air, or dreams real again explains insanity.— |

36   Analysis *of pleasures of scenery.*—

There is absolute pleasure independent of imagination (as in *hearing* music) , this probably arises from (1) harmony of colours, & their *absolute* beauty (which is as real a cause as in music) from the splendour of light, especially coloured.—that light is a beautiful object one knows from seeing artificial lights in the

37   night.—from the mere exercise of the | organ of sight, which is common to every kind of view—as likewise is novelty of view even old one. every time one looks at it.—these two causes very weak. (2^d) form. Some forms seem instinctively beautiful /as round ones/;—then there the pleasure of perspective, which cannot be doubted if we look at buildings, even ugly ones.—the pleasure

from perspective is derived in a river from seeing how the serpentine lines narrow in the distance.—& even on paper two waving *perfectly parallel* lines are elegant.— |

38   Again there is beauty in rhythm & symmetry, of form—the beauty of [illegible] as Norfolk Isd fir shows this, or sea weed, etc., etc.—this gives beauty to a single tree,—& the leaves of the foreground either owe their beauty to absolute form or to the repetition of similar forms as in angular leaves,—(this Rhythmical beauty is shown by Humboldt from occurrence in Mexican &

39   Graecian to be single cause) [25] this symmetry & rhythm applies | to the view as a whole.[26]—Colour /and light/ has very much to do, as may be known by autumn, on *clear* day.—3ᵈ pleasure association *warmth, exercise,* birds *singings.*—

4ᵗʰ Pleasure of imagination, which correspond to those awakened during music.—connection with poetry, abundance, fertility, rustic life, virtuous happiness:—recall scraps of poetry;—former thoughts, & experienced people recall pictures & therefore imagin-

40   ing pleasure | of imitation come into play.—the train of thoughts vary no doubt in different people, an agriculturist in whose mind supply of food was evasive & ill-defined thought would receive pleasure from thinking of the fertility.—I a geologist, have ill-defined notion of land covered with ocean, former animals, slow force cracking surface etc truly poetical. (V. Wordsworth about

41   sciences being sufficiently habitual to become poetical.) | the botanist might so view plants & trees.—I am sure I remember my pleasure in Kensington Gardens has often been greatly excited by looking at trees at [i.e., as] great compound animals[27] united by wonderful & mysterious manner.—There is much imagination in every view. if one were admiring one in India, & a tiger stalked across the plains, how one's feelings would be excited, & how the scenery would rise. Deer in Parks ditto.— |

42   My Father says there is case on record he believes in Philosoph. Transactions of ideot 18 year old eating white lead. who was most violently purged /believe worms were passed off/ & vomited, but who when he recovered was found to be ignorant, but quite sensible & no ways an ideot.— ((In this case must have been functional.—)) He has some idea of a son of Dr. Priestly who was cured from a fall of ideotcy.—

The story of the Corbets & big noses quite conjectural, in Blakeways book of Sheriffs.— |

43   July 22ᵈ· 1838

No Deliriums, yet in some inflammatory diseases, when there has been no cloud on the mind, every occurrence for a day or two are absolutely forgotten.—My father signed a bond, yet when he paid the attorneys bill, he asked what bond he could have had. yet during whole illness, he had been able to direct about his own health.—His complaint was carbuncle on <head> neck.—He has seen other cases of similar nature.— like FitzRoy[28] in sleep giving

44   directions,—& forgetfulness | after bad accidents:—after journey, a fit of gout, has affected his memory of everything in journey. short time previous,—because, pain prevents repetition of idea.—

Mr. Blakeway has mentioned in Antiquities of Shrewsbury, something about big noses & name Corbet, perhaps nonsense,— look to it.

My father has somewhere heard (Hunter?) that pulse of new born babies of labouring classes are slower than those of Gentle-

45   folks, & that | peculiarities of form in trades (as sailor, tailor, blacksmiths?) are likewise hereditary, & therefore that their children have some little advantage in these trades.—[29]

Delirium seems to rest the sensorium.—analogous to sleep; some doctors care it by stimulus & afterwards patient sinks.— |

46   When a muscle is moved very often, the motion becomes habitual & involuntary.—when a thought is thought very often it becomes habitual & involuntary,—that is involuntary memory, as in sleep.—a new thought arises?? compounded of the involuntary thoughts.—An intentionally recollection of anything is solely by association, & association is probably a physical effect of brain, the /similar recurrent/ thoughts being function of same part of brain, or the tendency to habit of producing a train of thought.— |

47–48e   [excised, not found]

49   Fox[30] believes cats discover birds nests & watch them till the young are big enough to eat.—There was blackbird's nest, near hot house at Shrewsbury, which the cat was seen by Hubbersty[31] to visit daily to see how the young got on.—This nest the cat could

If cats will /ever/ eat little birds, this most curious instance of reason & abstinence.—[32] |

50   My Father remarks that things of great importance are easily forgotten, (if unconnected with fear etc) because people think that the importance of the event by itself will make it to be remembered. whereas it is the importance.—people very often forget where money is placed.— (How often one forgets where put one key, when all keys are placed) Memory cannot solely be number of times repeated, because some people can remember poetry when once read over.— |

51   The extreme pleasure children show in the naughtiness of bothering children shows that sympathy is based as Burke maintains on pleasure in beholding the misfortunes of others.—[33]

In young children, the violent passions they go into, shows how truly an instinctive feeling.

In reflecting over an insane feeling of anger which came over me, when listening one evening when *tired* ((how true the heart the scene of anger)) to the pianoforte, it seemed solely to be feel-

52   ings of discomfort, especially about heart as if | excited action accompanied violent movement; may not passion be the feeling ((consequent on the violent muscular exertion)) which accompanies violent attack.—Even the worm when trod upon turneth, here probably there is no feeling of passion. but muscular exertion consequent on the injury & consequently excited action of heart.— now this is the oldest inherited & therefore remains, when the actual movement does not take place.—A start is HABITUAL movement to avoid any danger.—Fear, shamming death, or running |

53   away. accompanied with want of muscular exertion, palpitation, voiding urine because done by some animals in defence, etc

Starting must be habitual /involuntary/ movement from wish to avoid some danger—but it is instinctive because Nancy[34] tells me very young babies start at anything they hear or see which frightens them.—Now every animal moves quickly away from any sudden sound or noise, & therefore brain has been accustomed to send a mandate to the muscles & when the noise comes it cannot help doing it.— ((Fanny Hensleigh[35] doubts whether young babies start.)) ((If children wink it is instinct.))

54   Fear must be simple instinctive feeling: I have awakened | in the night being slightly unwell & felt so much afraid though my reason was laughing & told me there was nothing, & tried to seize hold of objects to be frightened at— (again diseases of the heart are accompanied by much involuntary fear) In these cases probably the system is affected, & by *habit* the mind tries to fix upon some object:—When a man, child or colt has once been frightened & started much more apt, this partly owing to heart? readily taking same movements, senses being on the look out, & the convey-

55   ing means | from the senses to the mind being more alive.—How is it, with people nervous from illness, it must be an excited action in the involuntary mind which is startled.—

My Father says he should think that in old people, in their dotage, who sing the songs /& tales/ of infancy, it is very doubtful whether they could recollect these same things from any effort of will whilst their minds were sound. |

56   Caroline tells me that Nina[36] when brought from Shrewsbury to Clayton[37] (though so fond of her & of servant of Richard & of Mary & her bed brought from Shrewsbury) yet for a fortnight continued wretchedly unhappy, constantly whined, would not remain quiet in any room, would not sleep at night even when in bedroom—grew very thin, would not go out of house except with Caroline—After fortnight, continued to grow thin & did not seem

quite happy. in five weeks was so thin, that she was sent back to Shrewsbury, then immediately fell into her old ways & became fat! What remarkable affection to a place. How like strong feelings of man.— |

57   The sensation of fear is accompanied by /troubled/ beating of heart, sweat, trembling of muscles, are not these effects of violent running away, & must not <this> /running away/ have been usual effects of fear.—the state of collapse may be imitation of death, which many animals put on.—The flush which accompanies passion, & not sweat, is the stated [?] effect of short, but violent action.—[38]

To avoid stating how far, I believe, in Materialism, say only that emotions, instincts degrees of talent, which are hereditary are so because brain of child resembles parent stock.— (& phrenologists state that brain alters) |

58   It is known that birds learn to sing & do not acquire it instinctively. May not this be connected with their power of acquiring language.—

Hensleigh W.[39] says that babies know a frown very early in life. (I think I have seen same thing before they could understand what frowning means.) if so this is precisely analogous or identical with bird knowing a cat, the first it sees it.—it is frightened without knowing why—the child dislikes the frown[40] without knowing

59   why—a man | as in Guy Mannering.[41] feels pleasure in seeing the scenes of his childhood without knowing why—had not [been] conscious of recollecting it—this may be nearest approach to such instincts which full grown men can experience—

Instinctive walking of animals. that is the ready movement & co-relation of the proper muscles. may be illustrated by the extreme difficulty of moving muscles in different way from what they have been accustomed to in certain actions—the difficulty of getting on a horse on left side (not good example) because leg is right handed.— |

60   In Review (Edinburgh) [42] of Froude's life. that author remarks that writing down his confessions of sins did not make him more humble.—it has obscurely occurred to me that Capt. F. R. [i.e., FitzRoy] candour and ready confession of error made him less repentant.—In making too much profession, or rather in only *fully* expressing momentary feelings of gratitude, I had a sort of consciousness I was not right; though I never realized the idea that I was tending to make myself in *act* less grateful.—How comes this tendency in these cases? How did my mind feel it was wrong (&

61e it was not | merely morally wrong, but hurting my character I felt it—this is kind of conscience, in obscure memory of having read or thought of some such remark as now advanced: for I caught it like a flash.—strange if judgment remains, where reason is forgotten, it is conscience, or instinct.

Hensleigh says to say *Brain* per se thinks is nonsense; yet who will venture to say germ within egg, cannot think—as well as animal born with instinctive knowledge, but if so, yet this knowledge acquired by senses,—then thinking consists of sensation of 62e images before your eyes, or ears (language mere means | of exciting association) or of memory of such sensation, & memory is repetition of whatever takes place in brain, when sensation is perceived.—

Aug. 7th—38. Transaction of the Entomological Society of London Vol. I, p. 106—Col. Sykes[43] on *Formica indefessa* placed table in cups of water which they waded or swam across.—they then stretched themselves from wall to table—table being removed a little further, they ascended about a foot & leapt across (Col Sykes compares this with pidgeons finding their way home—there is something wrong in comparing these cases, when agency is un- 63e known, with simple exertion of | intellectual faculty) if ants had at once made this leap it would have been instinctive, seeing that time is lost & endeavours made must be experience & intellect.—

do. p. 157 Westwood[44] remarks that some imported plants are attacked by insects & snails of this country (thus Dahlias by snails) —*Apion radiolum* undergoes transformation in the stem of hollyhock, although ordinary Habitat is *Malva sylvestris*.

do. p. 228 Newport[45] says Dr. Darwin[46] mistaken in saying common wasp cuts off wings of flies from intellect, but it does it always instinctively or habitually.—good Heavens is it disputed that a wasp has this much intellect, yet habit may make it act wrong, as I have done when taking lid off <right> side of tea chest, when no tea. |

64e   do. p. 233 Mr Lewis[47] describes case of insects /a Perge/ of Terebrantia laying eggs on leaves of Eucalyptus, watching few days till larva excluded, then though not feeding them /nor helping larva from egg/ watching them, brooding over them, preserving them from the sun & enemies—would not fly away, but bit pencil when touched with it—do not know their own larvae, but one female may be moved to other larvae, when two groups near mother desert one sometimes & go to other, so that two mothers to one group— (as in birds blind storge—they continue till death. Thus acting 4 to 6 weeks. The deserted broods appeared healthy—This remarkable case may be normal with insects, but habit forgotten in all older species. The earwig & a doubtful one of Acanthosoma grisea described |

65–68e [excised, not found]

69 . . . as first caused by will of Gods. /or God/ secondly that these
are replaced by metaphysical abstractions, such as plastic virtue,
etc. (Very true, no doubt savage attribute thunder & lightning to
Gods anger.— (∴ more poetry in that state of mind: The
Chileno[48] says the mountains are as God made them,—next step
plastic <virtue> natures accounting for fossils). & lastly the tracing facts to laws without any attempt to know their nature. Reviewer considers this profoundly true.—How is it with children.—

Now it is not a little remarkable that the fixed laws of nature
should be /universally/ thought to be the *will* of a superior being,
70 whose natures can only be rudely traced out. When one sees | this,
one suspects that our will may /arise from/ as fixed laws of organization.—M. le Comte argues against all contrivance—it is what my
views tend to.—

When a man is in a passion he put himself stiff, & walks hard.
((He cannot avoid sending will of actions to muscles any more
than prevent heart beat.)) remember how Pincher does just the
same. I noticed this by perceiving myself skipping when wanting
not to feel angry—Such efforts prevent anger, but observing eyes
thus unconsciously discover struggle of feeling. It is as much effort
to walk then lightly as to endeavour to stop heart beating: on
ceasing, so does other. |

71 What an animal like taste of, likes smell of, ∴ Hyaena likes
smell of that fatty substance it scrapes off its bottom. it is relic of
same thing that makes one dog smell posterior at another.—

Why do bulls & horses, animals of different orders turn up their
nostrils when excited by love? Stallion licking udders of mare
strictly analogous to men's affect for women's breasts. ∴ Dr. Darwin's[49] theory probably wrong, otherwise horses would have idea
of beautiful forms.— |

72 With respect to free will, seeing a puppy playing cannot doubt
that they have free will, if so all animals, then an oyster has & a
polype (& a plant in some senses, perhaps, though from not having
pain or pleasure, actions unavoidable & only to be changed by
habits). Now free will of oyster, one can fancy to be direct effect
of organization, by the capacities its senses give it of pain or pleasure. If so free will is to mind, what chance is to matter /M. Le
73 Compte/—the free will (if so called) makes change | in bodily or-
ganization of oyster, so may free will make change in man.—the
real argument fixes on hereditary disposition & instincts.—Put it
so.—Probably some error in argument, should be grateful if it
were pointed out. My wish to improve my temper, what does it
arise from, but organization, that organization may have been af-

fected by circumstances & education & by the choice which at that time organization gave me to will—Verily the faults of the fathers, corporeal & bodily, are visited upon the children.— |

74 The above views would make a man a predestinarian of a new kind, because he would tend to be an atheist. Man thus believing, would more earnestly pray "deliver us from temptation," he would be most humble, he would strive <to do good> /to improve his organization/ for his children's sake & for the effect of his example on others. It may be doubted whether a man intentionally can wag his finger from real caprice. it is chance which way it will be, but yet it is settled by reason.— |

75 How slow habits are changed may be inferred from expression "relict of bad habit." as child is cured of sucking [?] his fingers by rubbing them with alum, so more slowly does animal leave off instinct, when attended with bad effects

Martineau.[50] How to observe, p. 21–26, argues /with examples/ very justly there is no universal moral sense /from difference of actions of approved/—yet as, I think, the opposite side has been shown—see Mackintosh,[51]—must grant that the conscience varies in different races. No more wonderful than dogs should have dif-

76 ferent instincts. Fact most opposed to this view where | the moral sense seems to have changed suddenly—but are not such /sudden/ changes rare,—as when Polynesian Mothers ceased to destroy their offspring? Yet perhaps if they had [illegible] their children, this moral sense, would have been as much, as in other races of mankind.

p. 27. Mart[ineau][52] allows *some* universal feelings of right & wrong /& therefore in *fact* only *limits* moral sense/ which she seems to think are to make others happy & wrong to injure them without temptation. This probably is natural consequence of man, like deer, etc., being social animal, & this conscience or in-

77 stinct may be | most firmly fixed, but it will not prevent others being engrafted. No one doubts patriotism & family pride are hereditary, & therefore he has these strong, & does not act up to them, no doubt disobeys & hurts conscience more than other.—A Scotchman will his country or Swis.—it may be answered effects of education, [but this] may be opposed [because there are] undoubted cases of hereditary pride & in single families. |

78 Edinburgh Phil. Transact. p. 365.[53] Case of double consciousness, one only /little/ less perfect than other, absolutely two people. Consider this profoundly, may throw light on *consciousness,* explained by Dr. Dewar[54] on principle of association ((fully bears out my Father's doctrine about people forgetting their insanity)).—there seem other cases somewhat analogous, & which I

think will lead to fact of old people singing songs of their child-hood, & certainly of Miss Cogan, & fully corroborates the fact of her not repeating song when she had recollected it in perfect senses.— These things, & drunkedness, show what trains of thought depend

79 on state of turn | In drunkedness same disposition recurs, such as . . . of Trinity always thinking people were calling him a bastard when drunk.—having really been so.—some always sentimental, some quarrelsome as B^e.[55] on board Beagle, some more good hu-moured as self.— ((When Miss Cogan has remembered her song, then the song was to her like one which though learnt in infancy, had often been repeated. Now it is remarked that A. Bessy re-peated things, which none about had EVER before heard, so very probably forgotten.)) Such facts bear on such characters as Al-len W.[56] & Babington,[57] both half ideotic in some respects, & with store of accurate & even profound knowledge or other unusual lines—both odd appearance about eyes.—One botanist & great knowledge of Irish Politics /both bad jokers/.—the other army officer, horticulture & religious sects, yet Allen W. remark about his slippers bad for fires. What is wrong in his head. & Babington's silly joking |

80 The possibility of the brain having whole train of thoughts, feeling & perception separate from the ordinary state of mind, is probably analogous to the double individuality implied by habit, when one acts unconsciously with respect to more energetic self, & likewise one forgets what one performs habitually.—Agrees with insanity, as in Dr. Ash's case, when he struggled as it were with a second & unreasonable man.—If one could remember all ones fathers actions, as one does those in second childhood, or when drunk, they would not be more different, & yet they would make one's father & self one person—& thus eternal punishment ex-plained. |

81 These facts showing what a train of thought, action, etc., will arise from physical action on the brain, renders much less wonder-ful the instincts of animals.—

Aug. 12th. 38. At the Athenaeum Club was very much struck with intense an headache /after good days work/ which came on from reading (review of) M. Comte Phil. which made me /en-deavour to/ remember, & to think deeply, & the immediate man-ner in which my head got well when reading article by Boz.[58]— Now in 'this I was interested as was I in the other, & read so in-tently as to be unconscious of all around, yet there was no strain on the intellectual powers—the difference is of a man wagging his foot, & working with his toe to perform some difficult task.— |

82 Aug. 12th. When in National Institution & not feeling much

enthusiasm, happened to go close to one & smelt the peculiar smell of Picture. association with much pleasure immediately thrilled across me, bringing up old indistinct ideas of FitzWilliam Museum.[59] I was amused at this after seven years interval.

Augt. 15th. As child gains habit /or trick/ so much more easily than man, so may animal obtain it far more easily, in proportion to variableness or power of intellect.—Some complicated trades can *hardly* be considered as actions otherwise than habitual.—Instances?? |

83e   The possibility of two quite separate trains going on in the mind as in double consciousness may really explain what habit is. In the *habitual* train of thought one idea calls up other & the consciousness of double individual is not awakened.—The habitual individual remembers things done in the other habitual state because it will (without double consciousness?) change its habits.— |

83   Aug. 16th. As instance of hereditary mind. I a Darwin & take after my Father in heraldic principle, & Eras a Wedgwood in many respects & some of Aunt Sarahs[60] cranks, & so is Catherine in some respects—. good instances when education same.—My handwriting same as Grandfather.—[61] |

84e   Aug. 16th Anger /Rage/ in worst form is described by Spenser (Faery Queene (Describt, of Queen) O[62] of Hell Cant IV or V) as *pale & trembling* & not as flushing & with muscles rigid.—How is this? (dealt with p. 241) [63] |

84   Origin of man now proved.—Metaphysics must flourish.—He who understand baboon would do more toward metaphysics than Locke

A dog *whines*, & so does man.—dogs laugh for joy, so does dog bark (not shout) when opening his mouth in romps, he smiles. Many of actions, as hiccough & yawn are probably merely coor-
85   ganic as | connexion of mammae & womb.—We need not feel so much surprise at male animals smelling vagina of females.—when it is recollected that smell of one's own pud not disagree[able.]— Ourang outang at Zoology Gardens touched pud of young males & smells its fingers. Seeing a dog & horse & man yawn, makes me feel how <much> all animals <are> built on one structure.

He who doubts about National character let him compare the American whether in the cold regions of the North,—the elevated
86   table land of Peru, | the hot plains of the Amazons & Brazil, with the Negros of Africa (or again the black man of Van Diemen's land & the energetic copper coloured natives of New Zealand) — the American in Brazil is under same conditions as Negro on the other side of the Atlantic. Why then is he so different in organi-
87   zation.— | same cause as colour & shape & ideosyncracy.—Look at

the Indian in slavery & look at the Negro—look at them both savage—look at them both semi-civilized—

Perhaps one cause of the intense labour of *original inventive* thought is that none of the ideas are habitual, nor recalled by obvious associations. as by reading a book.—Consider this— |

88   "The fledge-dove knows the prowlers of the air" etc etc etc so is conscience etc etc Coleridge,—Zapoyla p. 117, Galignani Edition.[64]

Fine poetry, or a strain of music, when the mind is rendered ductile by grief, or by bodily weakness, melts into tears, with sensations of sorrowful delight, very like best feeling of sympathy.— Mem: Burke's[65] idea of Sympathy. being real pleasure at pain of

89   others, with rational | [illegible] to assist them,—otherwise as he remarks sympathy would be barrier & lead people from scenes of distress,—see how a crowd collects at an accident,—children with other children naughty.—Why does person cry for joy?

17th. August. Montaigne Vol. I[66] has well observed one does not fear death from its pain, but one only *fears* that pain which is connected with death!—How has this instinctive fear arisen?

19th. When I went down to Woollich[67] I was trying to unbend my mind as much as possible (tasting success by decreasing headache) & found best plan was allowing my mind to skip from sub-

90   ject | to subject as quick as it chose, although thinking (& talking) for the moment with interest on each.—∴ my father is right in saying delirium rest—therefore dreams thus act.—∴ weak minded people are fickle & full of levity (?do I not confound action & thought here?) The opposite extreme of this desultory thought is following out such an idea, as effect of sea on coves[68] when waters had fallen, as in my Glen Roy paper. this greatest mental effort of which I am capable—. I suspect from these facts that

91   whole effort consists in keeping one idea before your | mind steadily, & not merely thinking intently, for that one does with novel for a length of time. Then if one endeavour to keep any simple idea—as scarlet steady before mind for period /if this scarlet were before one [it would be] effortless/ one is obliged to repeat the word, & think of qualities as flowers, cloth, etc. & with all this [it is] difficult EXPERIMENTIZE upon this effort.—it looks so analogous to muscle in one position great fatigue.—May explain excessive labour of inventive thought.—Examine frame of mind in follow-

92   ing changes during fall of sea.—Is the effort | greater if the idea is abstract as love (or an emotion; *not so*) than if simple idea as scarlet?—How can people dwell on pain? No definite idea. nor is an emotion—People who can multiply large numbers in their head must have this high faculty, yet not clever people.

Aug. 21st 38.

When a dog in play has his mouth open ready to bark, & lip

twisted up in that peculiar manner they do, even more than in a real snarl, they are enjoying a satirical laugh.—when snarling, real bitter sarcasm.— |

93   Seeing how ancient these expressions are, it is no wonder that they are so difficult to conceal. A man /insulted/ may forgive his enemy & not wish to strike him, but he will find it far more difficult to to look tranquil.—He may despise a man & say nothing, but without a most distinct will, he will find it hard to keep his lip from stiffening over his canine teeth.—[69] He may feel satisfied with himself, & though dreading to say so, his step will grow erect & stiff like that of turkey.—he may be amused, he need not express it, he may most earnestly wish [not] to do it, but an involuntary

94   laugh will burst forth. This & yawning (common to other | animals) , scream of agony, sigh of discomfort & weariness & meditative tranquility ((whine of children, puppies do [ditto], so dogs nearly silent, so with men.— How is crying—peculiar not common?) — no bark of anger nor have monkeys & many other animals,—but yet when angry it is hard not to growl out some sound even if it be inarticulate. the maniac shouts & bellows with passion.—It is not a little remarkable that those sounds which are involuntary, are common to animals.—Curious to trace which of these actions are habitual, & which now connected physical relations, like sighing to relieve circulation after stillness.—Now I conceive if organiza-

95   tion were changed, I conceive sighing might | yet remain just like sneering does.—is yawning habitual from awakening from sleep see how a dog yawns when he awakes. & stretching & yawning can be explained from too long rest of muscles, evidently habitual when transferred, (also how often) to the tale of a wearisome man.—

Is frowning result of straining vision, as savages without hats put up their hands, & as attention would amongst lowest savages clearly be directed chiefly by objects of vision.—

Does the contraction & wrinkling of skin contract iris?—same way as one lifts up eyebrows to see things in dark, & hence is this the cause of expression of surprise, viz seeing something obscurely with the wish to make it out?— |

96   Seeing a Baby (like Hensleigh's) [70] smile & frown, who can doubt these are instinctive—child does not sneer, because no young animal has canine teeth.—A dog when he barks puts his lips in peculiar position, & he holds them this way when opening mouth between interval of barking, now this is smile.

With respect to sneering, the very essence of an habitual movement is continuing it when useless,—therefore it is here continued when the uncovering the canine useless.—

The distinction /as often said/ of language in man is very

97   great from | all animals—but do not overrate—animals communi-
cate to each other. Lonsdale's[71] story of Snails, Fox of cows, &
many of insects—they likewise must understand each other's expressions, sounds, & signal movements.—some say dogs understand
expression of man's face.—How far they communicate not easy to
know,—but this capability of understanding language is considerable. Thus carthorse & dogs.—birds many cries. monkeys
communicate much to each other.— |

98   Waterhouse[72] says far more instincts in ALL of the Hymenop-
tera; than in other orders (study Fish with this view) therefore
there is Instinctual developement in one order, as there is Intellectual in human—probably some genera in different orders more
advanced than others just as dog & Elephant most intellectual.—
Hymenoptera typical insects, i.e., have all parts. Waterhouse |

99   Study well the greater number of insects in insecta—not connected with transformation because Spiders have many—great
powers of communicating knowledge to each other—

August 23ᵈ. Jones[73] said the great calculators, from the confined nature of their associations (is it not so in punning) are
people of very limited intellects, & in the same way are chess
Players. A man at Cambridge, during his time almost an absolute
fool used to play regularly with D'Arblay[74] of Christ of *great
100 genius,* & yet invariably used to beat | him—The son of a Fruiterer
in Bond St. was so great a fool that his Father only left him a
guinea a week, yet he was inimitable chess player.—Peacock's[75]
remarks about mathematicians not being profound reasoners.—
all same fact—for, as Jones observed, in playing chess however
many places & contingency a man has [to] keep in mind, all is certain—there is judgment of probabilities, therefore this judgment
gives a man common sense, & the highest intellectual powers of
101e perceiving | & classifying <distinct> resemblances.— (Can't go
with this FD) [76]

The facts of half instincts when two varieties are crossed as in
Shepherd dogs /Inherited habits: Horse/[77]—is valuable it shows
that new instincts can originate.—strong argument for brain bringing thought, & not merely instinct, a separate thing superadded.—
we can thus trace causation of thought.—it is brought within
limits of examination.—obeys same laws as other parts of structure.[78]

Can an analogy be drawn /hereditary/ associated pleasures &
pains & emotions—such as child sucking, gives pleasure, & always
has done therefore sight of own child (when frame [illegible] in
condition to receive pleasure) gives pleasure, ie., love.—& so pain
gives fear of death.[79] |

102e Mayo Philosophy of Living[80] p. 140—Dreams good account of
/thinks/ are recollected when intense, or when so near waking,
that an associate is kept up with waking thoughts.—Ld Brougham
thinks no dreams except at this time! how does he account for
dogs & men speaking in their sleep.—Characters of dreams no surprise, at the violation of all <rules> relations of time <identity,>
place, & personal connections—ideas are strung together in manner quite different from when awake.—peculiar sensation as flying.
(No memory of past events?) or influence on our conduct the links

103 which when conscious connect past, present & future | thoughts are
broken.—Sir J. Franklin[81] when *starved.* all party dreamt of feasts
of good food.—The mind wills to do this, & hears that, but yet
scarcely really moves.—the willing therefore is ideal, as all the
other perceptions.—

The mind thinks with extraordinary rapidity—We may conclude that neither number, vividness, rapidity, novelty of separate
ideas cause fatigue to the mind, it is solely the comparison, with
past ideas, which makes consciousness—& which tells one of
reality—castle in the air is more prolonged than dream, hence
fatiguing,—else it is only our consciousness, & senses tell us it is
not real.—dreaming appear clearly rest of the mind. With all
other faculties /∴ Vide page 110 by mistake./[82] |

104 N.B. Everything which happens to man who does not produce
children, or after he has useless, does not affect man. argument for
early education.—fear of death!!! as Montaigne observes, distinct
from pain, for one hates pain from this fear—& not death for the
pain.—How was this instinct gained? By conversation—∴ modified
in those races where it is customary to die—

August 24.th. As some *impressions* (Hume) [83] become unconscious,
so may some *ideas,* i.e., habits, which must require idea to order

105 muscles to do the action. ?is | it the impression becoming *very
often* unconscious, which makes the *idea* unconscious, if so (think
of this) study what impressions become unconscious those which
are viewed with little interest, & those which are viewed very often.
—former do not give rise to ideas so much as objects of interest.—
do. I was much struck with observing how the Baboon /<Ma-
caco> Cyanocephalus Sphynx Linnaeus/ constantly moved the
skin of forehead over eyes, at every motion & look /turn/ of the

106 head. | I could not perceive /any/ distinct *wrinkle,* but such
movements in skin of eyebrow important analogy with man.—I see
monkeys *grin* with passion, that is show all the teeth 〈(and make
noise not like pish, but like chit-chit-chit, quickly uncovering their
teeth, this the Keeper thinks is from pleasure: may be compared
to laughing)〉. they dance with passion, ie. nervous impulse to

action is sent so fast to limbs that they cannot remain still.—I do not doubt this Baboon knew women.—Another little old American monkey /Mycelis/ I gave nut, but held it between fingers, the
107 peevish expression was | most curious. remember the expostulatory angry look of black spider monkey when touched, also another monkey to dog. I showed the nut & then closed my [hand] Mem. expression of fury, jump to scratch my face. The ourang outang, under same circumstances, threw itself down on its back & kicked & cryed like naughty child.—do monkeys cry?—((They *whine* like children.))

Expression is an hereditary habitual movement consequent on some action, which the progenitor did when excited or disturbed by the same cause, which /now/ excites the expression.— |

108   Habitual actions are the reverse of intellectual, there is no comparison of ideas—one follows other as in blindest memory—also low faculty of understanding.

Adam Smith[84] (D. Stewart life of p. 27) says <sympathy> we can only know what others think by putting ourselves in their situation & then we feel like them—hence sympathy very unsatisfactory because does not like Burke explain pleasure.

August 26th I cannot help thinking horses admire a wide prospect—The very superiority of man perhaps depends on the num-
109 ber of sources of pleasure & innate tastes, he | partakes, taste for musical sound with birds & ?howling monkeys—smell with many animals—see how a dog likes smell of Partridge—Man's taste for smell of flowers owing to *parent* being fruit eater.—origin of colours?—

Nothing shows one how little happiness depends on the senses [more] than the <small> fact that no one, looking back to his life, would say how many good dinners or . . . . . he had had; he would say how many happy days he spent in such a place.— |
110   (Vide page 103, supra, by mistake.)

have lower animals these vivid thoughts In same book[85] (p. 143) wonderful case of perfect double consciousness Mayo compares it with somnambulism—the young lady almost equally in her senses in either state—does this throw light on instinct, showing what trains of action may be done unconsciously as far as the ordinary state is concerned?—

Mr. Mayo told me that case of a lady (whose name was told me who told the fact to Mr. Mayo himself.[86] she was one day reading a book, with ivory paper cutter, which she valued, & she was suddenly called to go on the lawn to see something, on her return
111 could not find paper cutter. hunted in vain | for it—ten years afterwards whilst at a meal, she suddenly like a flash without any

assignable cause, remembered she had put it in branch of tree, & apologizing to party, went out & found it there!!! Lady in perfect /mental/ health.— ((Erasmus had almost same thing happen to him about a knife which he had hid some years before—was greatly astonished at the time & could trace no chain of association))

Mayo Philos. seems certain that muscular, mental, digestive nervous influence replace each other

August 29th. Went to Bed & built /common/ Castle in the air, of being compelled, from some quite imaginary cause to start at once to Shrewsbury, vaguely thought of packing up.—was lying on my back fell to sleep for second & wakened.—had very clear & pretty vivid /& perfectly characterized/ dream, in continuation of waking thought—my servant was in the room, with my trunk out & I was engaged in hurriedly giving orders.—Now what was

112 difference between Castle & dream | No answer shows our profound ignorance in so simple case.—There was memory, for it related to past idea.—there was a kind of ideot consciousness for moment, implied by /presence/ my servant, /box,/ my own manner of ordering things to be done.—The senses are closed probably by sleep & not vice versa. anyhow I might have been quite still, & not attending to bodily sensations & yet the Castle would not have turned into dream.—It appears to me, that the mind is wholly absorbed with *one* idea (hence *apparent* vividness) & there being no other parallel trains of ideas connected with past circumstances, as whether I really was going to Shrewsbury, whether I had rung for Covington,[87] whether he had come & opened box,

113 whether I had thought what clothes to take (how often | one cannot tell whether one has rung the bell, when one recollects circumstances were such one naturally would do so!) Now all these parallel trains of thought necessary heirs of every action, & always running on in mind, being absent. one could not compare the castle with them. therefore could not *doubt* or *believe*. When I say trains, it may be instantaneous changes in order calling up ideas of every late impression.— (do the ideas, direct effect of perception by senses fail first, as whether I had pulled the bell??) .— It may be deception to say the mind <thinks> quicker in sleep, it may do less work & yet do so, from the exertion of keeping up the memory of every late impression. & likewise gaining new ones from senses. & <comparing them> calling up old ones, to be sure of one's consciousness. |

114 Mayo[88] observes no improbabilities in a dream, effect of doubting nor believing, effect of not reasoning, effect of not having other trains of thought, or memory from innumerable late events.

—the fatigue of thinking is keeping up these trains, especially if they are inventive as in imagination & in rigidly comparing each step as in reasoning—hence delirium & sleep mental rest, though most vivid & rapid thought.—

There may be some /two or three/ trains of thought, therefore one may imperfectly reason—Abercrombie's[89] case of /in Botanical Student/ somnabulism, did reason about himself—but not about facts gained, or gaining by senses.—As [in] sleep only one idea is
115 awake, | when one is awake many necessarily are, when one is deeply reasoning besides these (which must be present, though one is not conscious of them, else one would not stand) a crowd of other trains of thought are in progress—In castle of air the *trouble* /I well recollect/ is in making things somewhat probable, in comparing every step, & inventing new means.—therefore works of imagination *hard* work.—Keeping one idea present is, perhaps, hard work—though dreams do that.

One Reflective Consciousness is curious problem, one does not care for the pains of ones infancy—one cannot bring it to one self. —nor of a bad dream, when that is not recollected, nor of the Botanical Somnambulist (if he had been unhappy) —it is because
116 in this | state, the consciousness does not go back to former periods so as to give one individuality in this case.—But now in Mayo's /p. 140/ case of double consciousness, one would pity suffering in one state almost as much as in the other,—though she when well did not recollect anything.—if one was subject to this disease oneself, one would only feel sympathy, as for for the heard suffering of a dear friend—this gives one strong idea of what individuality is.—Insanity is somewhat the same as double consciousness, as shown in the tendency to forget the insane ideas, & one's expres-
117 sion of | double self, though as in Dr. Ashe's case, one here was conscious of the two states.—

August 30th. It is singular when looking at a table one has vague idea something is not there, & then when one begins eating one perceives butter or salt is not there.—the reality does not resemble the picture in one's mind, but one does not stop to *reason* what there should be & *discover loss.*

Definition of happiness the number of pleasant ideas passing through mind in given time—intensity to degree of <happy> pleasure of such thoughts. |
118 We give no credit to instinctive feelings.—for man losing his children, any more than dog losing his puppies—This looks like free will.—

V. last page. A healthy child is /more/ entirely happy (contentment is different it refers to *wishes* for future) than perhaps well

/regulated/ philosopher—yet the philosopher has a much more intense happiness—so is it when same man is compared to peasant.—To make greatest number of pleasant thoughts, he must have
119 a contingency of good food, no pain,—& the sensual | enjoyment of the minute add[ed] to the happiness,—but as they are not recollected whether from frequency, or inherent structure of mind, they make, either in themselves, or if recollected, such part of thoughts innumerable, which past through mind.—Those thoughts are most pleasant, when the conscience tells our [mind] good has been done /& conscience free from offence/—pleasure of intellect affection excited, pleasure of imagination—therefore do these & be happy, & these pleasures are so very great, that every one who
120 has tasted them, will think | the sum total of happiness greater even if mixed with some pain,—than the happiness of a peasant, with whom sensual enjoyments of the minute make large portion of daily <happiness> pleasure. A wise man will try to obtain this happiness, though he sees some /intellectual/ good men, from insanity etc unhappy—perhaps not so much as they appear & perhaps
121 partly their fault.—Whether this rule of | happiness agrees with that of New Testament is other question.—little is there said of intellectual cultivation, main source of the intense happiness.—it is again another question, whether this happiness is the object of living,—or whether if we obey literally New Testament future life is almost the sole object.—I doubt whether the last be right. The
122 two rules come very near each other.— | The rules to mortify yourself do not tend to this—though *believing* it to be true, & *then acting* on it, will add to happiness.—

Men having some instincts as revenge /and anger/, which experience shows it must for his happiness to check—that is external circumstances are so conditioned as they are effecting a change in his instincts—like what is happening with other animals—is far
123 from odd | nor is it odd he should have had them.—with lesser intellect they might be necessary & no doubt were preservative, & are now, like all other structures slowly vanishing—the mind of man is no more perfect than instincts of animals to all & changing contingencies, or bodies of either.—Our descent, then, is the origin of our evil passions!!—The Devil under form of Baboon is our grandfather!— |

124 A man, who perfectly obeys his conscience or instinct, would probably feel but little that of anger or revenge.—they are incompatible & the former, the more pleasant.—

Simple happiness /as of child/ is large proportion of pleasant to unpleasant mental sensations in any given time /compared to what other people experience/.—But then sensation may be *more* or *less* pleasant & unpleasant, in same time,—therefore degrees of happiness—*Entire happiness* not being so desirable as <broken>

125 *intense* | happiness even with some pain, compared to what others experience in same time.—*Pleasure* more usually refers to the sensations when excited by impressions, & not mental or ideal ones, & these must occupy greater proportion of every man's time.—

Begin discussion—by saying what is Happiness?—When we look back to happy days, are they not those of which all our *recollections* are pleasant.— |

126   Browne Religio Medici, p. 21–24.[90] Curious passages showing how easily chance & will of Deity are confounded—well applicable to free will.

Mayo. Philosop. of Living, p. 293.[91] Animals "have notion of property" —their own property (—regarding food & in birds of place for nest) with dogs "have notion of master's property." ?Is not this rather more friendship—

Scott's Life, Vol. I, p. 127.[92] Talks of difficulty of his own drawing compared to a friend whose who[le] family can draw—says friend viewed him as Newfoundland dog would Greyhound about dread of water—INNATE. |

127   Septemb 1—If one performs some actions, which are pleasant, every concomitant circumstance calls up pleasure, or pleasure or pain of association.—now if one has these feelings, without being aware of their association /ie., hereditary/, does one not call them instinctive emotions?—

Dr. Holland[93] remarked that *insanity* like *sleep.* does not doubt the reality of the impression on its senses.—insane people believe they *hear* as well [as] see things which have no existence.—He compared spectral illusion[94] & insanity, the connexion appears to him vague—

Delirium of every degree of intensity ((In old man, he had just seen mind went on RAMBLING till excited by question.)) |

128   Sept. 4th.[95] Lyell[96] in his Principles talks of it as wonderful that Elephants understand contracts, but W. Fox's dogs that shut the door evidently did, for it did with far more alacrity when something good was shown him, than when merely ordered to do it.—

Plato /Erasmus/ says in Phaedo that our *"necessary ideas"* arise from the preexistence of the soul, are not derivable from experience.— read monkeys for preexistence. |

129   The young Ourang in /Zool./ Gardens *pouts.* partly out [of] displeasure (& partly out of I do not know what when it looked at the glass) when pouting protrudes its lips into point.—Man, though he does not pout, pushes out both lips in contempt, disgust & defiance.— different from sneer—

How easily horses associate sounds may be seen by omnibus's Horses starting, when door shut or cad cries out "right," or Drink-water's[97] horse jumping when word jump said—

I saw the ourang take up a stone & pound the earth.

Lockart's Life of W. Scott, Vol. VII, p. 35,[98] "as ideas come & the pulse rises, or as they flag & something like a snow-haze covers my whole imagination." |

130   September 3ᵈ Why when one thinks of any object (or naving looked at any object one shuts ones eyes) is the image not vivid as in sleep— (one can dream of intense scarlet??) is it because one then has no immediate comparison with perceptions, & that one fancies the image more vivid?—Surely the image in a dream cannot truly be as vivid /as reality [readily?]/ as in Spectral images.— |

131e´   Mem Chiloe Sow who carried from all parts straw to make its nest Pigs & Elephants (both Pachyderms) much intellect.—

mem: Yarrell's story of wheel horse in drays, scraping against cornice stone to cause friction

Athenaeum 1838 p. 652 Dʳ Daubeny[99] on the direction of mountain chains in N. America.

Fear probably is connected with *habitual* stopping of breath to hear any sound.—attitude of attention

((So intimately connected is passion with sending force to muscles, that in my grandfather remark, a tired man involuntarily feels angry, when brain is pumping force to legs & hands, & especially, when the whole body being paled [?], & not to any particular muscle)) |

132e   Sept 8ᵗʰ· I am tempted to say that those action which have been found necessary for long generation, (as friendship to fellow animals in social animals) are those which are good & consequently give pleasure, & not as Paleys[100] rule is then that on long run *will* do good.—alter *will* in all such cases to *have*, & *origin* as well as *rule* will be given.—

Mitchell Australia Vol. I, p. 292[101] "Dogs learn sooner to take kangaroos than emu, although *young* dogs get sadly torn in conflicts with the former. But it is one thing for a swift dog to over-

133–134e [excised, not found]

135 take an emu, & | notions are not effects of impressions long repeated, without the powers of the mind being EQUAL to the smallest casuistical doubts.—The history of Metaphysicks shows that such a view cannot be, anyhow, easily overturned.—So ready is change, from our idea of causation, to give a cause (& no one being apparent, one fixes on imaginary beings, many vicarious, like ourselves) that savages (Mem York Minster) [102] consider the thunder & lightning the direct will of the God (<thus> & hence

arises the *theological* age of science in every nation according to
136 M. le Comte) .[103] Those savages who thus | argue, make the same
mistake, more apparent however to us, as does that philosopher
who says the innate knowledge of creator <is> /has been/ implanted in us (?individually or in race?) by a separate act of God,
& not as a necessary integrant part of his most magnificent laws.
which we profane in thinking not capable to produce every effect
of every kind which surrounds us. Moreover /it would be difficult
to prove this/ this innate idea of God in civilized nations has not
been improved by culture ((who feels the most implicit faith that
through the goodness of God knowledge has been communicated
to us)). & that it does exist in different degrees in races.—whether
137 in Ancient Greeks, | with their mystical but sublime views, or the
wretched fears & strange superstitions of an Australian savage or
one of Tierra de Fuego.—

Mr. Miller (superintendent of the Zoological Gardens) remarked that the expressions & noises of monkeys go in groups.
Thus the pig-tailed baboon, shoved out its lip, looking absurdly
sulky /as/ often as keeper spoke to it,—but he thinks not sulkiness
—this expression he believes is common to that group—this is
very important as showing that expression mean SOMETHING.— |
138   Hunt (the intelligent Keeper) remarked that he had never seen
any of the *American Monkey* show any desire for women. ((Very
curious as they separate in structure)) ((The monkeys understand the affinities of man better than the boasted philosopher
himself)) —it is chiefly shown in old male.—A very green monkey
(from Senegal he thinks Callitrix Sebe??) he has seen place its
head downwards to look up women's petticoats, just like Jenny
with Tommy ourang. Very curious.—

Mr. Yarrell has seen Jenny, when Keeper was away, take her
chair & bang against the door to force it open, when she could
139enot succeed of herself.— | I saw <the male> Jenny untying a
very difficult knot—the sailor on board the ship could not puzzle
her—/Descent 1838/[104] with aid of teeth & hands.—It was very
curious to see her take bread from a visitor & before eating /every-
time/ look up to /keeper/ to see whether this was permitted &
eat it.—good case of association.— ((Listened with great attention
to Harmonicon, & readily put it, when guided to her own
mouth.— seemed to relish the smell of Verbena & Pocket Handker-
chief & liked the taste of Peppermint.—)) Perfect understand
voice—will do anything.—will take & give food to Tommy, or any-
140ething of any sort.—I saw Tommy picking his | nose with a straw:
Jenny will often do a thing which she had been told not to do.—
when she thinks keeper will not see her.—but then knows she has
done wrong & will hide herself.—I do not know whether fear or

shame.—When she thinks she is going to be whipped, will cover herself with straw, or with a blanket.—these cases of commonly using foreign bodies, for end, most important step in progres-
141 sion.— | The male Black Swan is very fierce when female is sitting, the keeper is obliged to go in with a stick, if he drops it, the bird will fly at him—Knowledge.—

Sept. 13th It will be good to give Abercrombie's[105] definition of "reason" & "reasoning," & take instance of Dray Horse going down hill (argue sophism of association, Kenyon) [106] & then go on to show, that if Cart horse argued from this into a theory of friction & gravity, it would be discoverer [of] "reasoning" or "reasoning," [sic]—only rather more steps.—dispute about words.— |

142 Miss Martineau (How to Observe p. 213) [107] says charity is found everywhere (is it not present with all associated animals?) I doubted it in Fuegians, till I remembered Bynoes[108] story of the women.—The Chillingham cattle[109] (& Porpoises) have not charity—is it in former case instinct to destroy contagious disease.— (Useful to use term instinct when origin of hereditary habit cannot be traced.)

V. D. p. 111,[110] case of Association.

Sept. 16th Zoological Gardens—Endeavoured to classify expressions of monkeys—I could only perceive that the American ones often put on a peevish expression, but not nearly so often <illegi-
143 ble> hardly ever the expression | of passion with open mouths like the old world ones.—Though they move whole skin of head, they do not move eyebrows. (I see some of the old world ones move skin of head & ears, ∴ *some* men have this power *abortive* muscles) The black Spider Monkey, very different disposition from others, slow cautious, angry cross look, followed by protrusion of lips, in which respect resembles some of the old ones. S. American group sneer—.

Sept. 21st Was witty in a dream in a confused manner. Thought that a person was hung & came to life, & then made many jokes about not having run away & having faced death like a hero, & then I had some confused idea of showing scar behind (instead of front) (having changed hanging into his head cut off) as kind of
144 wit showing he had | honourable wounds. all this was kind of wit. —I changed I believe from hanging to head cut off /there was the feeling of banter and joking/ because the whole train of Dr. Monro[111] experiment about hanging came before me showing impossibility of person recovering from hanging on account of blood, but all these ideas came one after other, without ever comparing them. I neither doubted them or *believed* them.—Believing consists in the comparison of ideas, connected with judgment.

[What is the Philosophy of Shame & Blushing?][CD112]

((Does Elephant know shame—dog knows triumph.—))

Sept. 23[d.]

Horses in omnibus *instantly* start when they hear ready, but if they see anything ahead which cad cannot see, they do not move muscle.—reason |

145e   The laughing noise which C. Sphynx made at Z. Gardens may be described as partaking of <st.> made by inspiration & quickly retracting tongue from behind upper & little between incisors.— like <noise which (when) > person says "what a pity"—

Lavater's Essays on Physiognomy translated by Holcroft /Vol. I/ p. 86[113] "We ought never to forget— —; that every man is born with a portion of phsiognominical sensation as certainly as every man who is not deformed is born with two eyes." I think this cannot be disputed anymore in man than in animals.—

In the drawings of Voltaire why is under lip curled over upper with mouth shut expressing *cool* irony, *not biting?*—[114] |

146e   What is analysis of expression of desire.—is there not protrusion of chin, like bulls & horses! ((1838))[115]—good instance of useless muscular tricks accompanying emotion.—when horses fighting, they put down ears, when kicking /turning round to kick/[116] they do the same, although it is then quite useless ((Cats Kneeling when old, like kittens at the breast))—now if horns were to grow on horses, they must yet continue to put down ears, when kicking.—good case of expression showing real affinity in face of donkey horse & zebra when going to kick.—Why does dog put down ears when pleased.—is it opposite movement to drawing them close on head, when going to fight, in which case expression

147   resembles a | fox—I can conceive the opposite muscles[117] would act, when in a passion.—dog tail curled when angry & very stiff. back arched. just contrary. when pleased tail loose & wagging—if as (I believe) Hunter says. neither fox nor wolf wag their tails, etc—it is very curious, recurrence of pleasure, so teaching expression /as constant smiles, cheerful face/.—Man when at ease has smooth brow contrary to wrinkled: (a horse when winnowing & pleased pricks his ears?) .—How is expression of anger in species of swans, in parrots etc etc.—peacock & turkey cock, in passion.— Cat when pleased, erects its tail & makes it very stiff. /& back/. when savage /no/ & ready to dash at prey streaked out & flaccid, when furious /with fright/ back absurdly arched & tail stiff.— |

148   —is shame, jealousy, envy all primitive feelings, no more to be analyzed than fear or anger? I should think shame would be more easily analysed than jealousy, because less discoverable in animals than latter.—Yet I think one can remonstrate with a dog, & make him ashamed of himself, in manner quite different from *fear;* there

is no inclination to jump away, it is ill-defined fear.—Yet one knows oneself it is quite different from that,—like /slight/ passion 149 from blood rushing in face, with less action of the heart.— | tendency to muscular movement. hence shy people (shame of ridicule) are singularly apt to catch tricks.—so are people in passion my F. rubbing hands,—stamping, grinding teeth.—in shame frowning, & anguish.—shyness not so,—affected laughter.—

A dog who goes home from shooting. runs away. is not afraid the whole way. but ashamed of himself.—Jealousy probably originally entirely sexual; first try to attract female (or object of attachment) & then failing to drive away rival.—

Fear is open mouthed to *hear,* though in individual case, nothing can be heard.—118 |

150 Shame would never make person tremble, like fear.—Why does any great mental aflection make body tremble? Why much laughter tears.—& shaking body—Are those parts of body, as heart, & chest (sobbing) which are most under great sympathetic nerve most subject to habit, as being less so [to] will.—

May not moral sense arise from our enlarged capacity /yet being obscurely guided/ <acting> or strong instinctive sexual, parental, & social instincts, giving rise "do unto others as yourself." "love 151 thy neighbour as thyself." Analyse this out, bearing | in mind many new relations from language.—The social instinct more than mere love.—fear for others acting in unison.—active assistance, etc., etc. it comes to Miss Martineaus119 one principle of charity.— ?May not idea of God arise from our confused idea of "ought," joined with necessary notion of "causation," in reference to this "ought," as well as the works of the whole world. Read Mackintosh120 on Moral Sense & emotions.—

The whole argument of expression more than any other point of structure takes its value from its connexion with mind, (to show hiatus in mind not saltus between man & Brutes) no one can doubt this connexion.—look at faces of people in different trades &c &c &c |

152 I observed the Asiatic Leopard *quarrelling,* mouth wide open, each [lip] drawn back & driving air out of mouth /wide open/ with prodigious force,—making growling, guggling noise /on back hairs erect/ Puma did same & some others—Thus <sudden> /forcible prolonged/ expulsion of air /dogs snarl much same way/ generic manifestation of great passion.—I do not think they arch their backs—Bengal tiger, when slightly angry, curls tip of tail.— do *two* cats arch their back when *fighting,* & not with dog, when fear might enter?—

I believe common Swan, arch raises neck & depresses chin,—

153estrikes with | wing arches wings—as does black Swan.—Goose do [ditto], all species put their necks straight out & hiss.—[Hyaena pisses from fear so does man.—& so dog]:CD121 Man grins & stamps with passion. Can expression be used more correctly than this for C. Sphynx.—In the wild ass there is a curious drawing out of the side part of nostrils, when passion commences.—

Nearly all will exclaim, your arguments are good but look at the immense difference between man, forget the use of language & judge only by what you see. Compare the Fuegian & Ourang-Outang, & dare to say differences so great . . "Ay Sir there is much in analogy we never find out." |

154e   This unwillingness to consider Creator as governing by laws is probable that as long as we consider each object an act of separate creation, we admire it more, because we can compare it to the standard of our own minds, which ceases to be the case when we consider the formation of laws invoking laws & giving rise at last even to the perception of a final cause. |

155   Read: Paper on consciousness in Brutes & Animals in Black-woods Magazine, June, 1838.122

(copied)

Mr. H. C. Watson123 on Geographical Distribution of British Plants.

A Volume published by Colonel124 in army on "Wheat" in Jersey—very curious facts about early production of foreign seeds.—many varieties. R. Jones125 has it—very curious book.—

Hume's126 essay on the Human Understanding well worth reading.

(copied).

D. Stewart127 <Smith> lives of Adam Smith. Read, etc. worth reading as giving abstract of Smith's views |

156   ((Take a pound of inflorescent parts of mosses & see if Hybrid can be made & ferns.))

Would a sensitive plant if irritated very regularly at one time every day naturally close at that time after long period—

My Father about double consciousness, & somnambulism.

Do people when inhaling nitrous oxide forget what they did when in this state, or remember what they did in former one.

about hereditary tricks & gestures, other cases like D. Corbet; do ideots form habits readily??

Do the Ourang Outang like smells /peppermint/ /& music/ Have the monkeys lice?—picture.—

Do female monkeys not show signs of impatience when women present?

Do they pout, or spit, or cry.—Shame, independent of fear: colour at base [of] nails—& of eyes.—

Do female monkeys care for men.—

Have we any ferns in the hothouse at home |

157   Has my father ever known <intemperance> disease in grand-child, when father has not had it, but where grandfather was the cause of his intemperance. <No.> Cannot say.—

Natural History of Babies

Do babies start (*i.e.,* useless sudden movement of muscles) very early in life Do they wink, when anything placed before their eyes, very young, before experience can have taught them to avoid danger Do they know frown when they first see it?

Charles Darwin
36 Great Marlborough St
PRIVATE

Notes by Paul H. Barrett

The task of tracing the particular books and articles used by Darwin has not always been easy, or possible. In some instances, editions printed later, or earlier, than those actually read by Darwin have been cited.

1. Sykes, W. H., "Descriptions of New Species of Indian Ants," *Transactions of the Entomological Society of London,* 1 (Pt. 2) :99–107, 1835.

2. Leighton, Francis Knyvett (1772–1834), Mayor of Shrewsbury, 1834, Ref.: Morris, Joseph, "The Provosts and Bailiffs of Shrewsbury," *Shropshire Archeological Society Transactions,* 3rd Ser., Vol. 5. See also Barlow, Nora, *Charles Darwin and the Voyage of the Beagle* (London: Pilot, 1945), p. 118: Col. Leighton's death is mentioned with regret in a letter April 23, 1835, from Charles Darwin to his sister Susan.

3. Probably Dryden Robert Corbet of Sundorne (1805–1859), son of John Corbet, M.P., Shrewsbury and High Sheriff of Salop (1793). Ref.: Burke, Bernard, *A Genealogy and Heraldic History of the Landed Gentry,* Burke, London, 1925, p. 395.

4. Darwin's interest in memories of early childhood may have prompted him to write, in 1838, when 29 years old, an autobiography beginning with his earliest recollections. See Darwin, Francis, and A. C. Seward, editors, *More Letters of Charles Darwin,* 2 vols., Murray, London, 1903, Vol. 1, pp. 1–5.

5. *Zoonomia,* p. 437: "Master A. about nine years old, had been seized at seven every morning for ten days with uncommon fits . . . he began to complain of pain about his navel, or more to the left side, and in a few minutes had exertions of his arms and legs like swimming. He then for half an hour hunted a pack of hounds; as appeared by his hallooing. . . ."

6. Scott, Walter, *The Antiquary,* 3 vols., Ballantyne, Constable, Edinburgh, 1816, Vol. 3, p. 220: ". . . shrill tremulous voice of Elspeth chaunting forth an old ballad in a wild and doleful recitative."

7. Aunt Bessy, i.e., Elizabeth Wedgwood (1764–1846), wife of Josiah Wedgwood of Maer, and mother of Emma, Darwin's wife.

8. For a genealogy of the Corbets of Shawbury Park, of Moreton Corbet, and of High Hatton Hall, see *The Family of Corbet: Its Life and Times,* St. Catherine Press, London [1915].

9. Train (of ideas). Erasmus Darwin in *Zoonomia,* often uses this expression, e.g., p. 46.

10. Kinlet, 25 miles southeast of Shrewsbury. Arthur Childe, born April 2, 1820, son of William Lacon Childe, Esq., M.P., of Kinlet Hall, was committed to the lunatic asylum in 1838, and again declared insane and recommitted by a Commission of Lunacy in 1854, on the basis of his having written 203 letters in code to the Queen expressing his infatuation for her. In his own defense Captain Childe, in his 43rd year, testified that the letters had been written for amusement and to keep himself occupied during his incarceration. Ref.: "Commission of Lunacy on Capt. Childe. A Lover of the Queen," *Shrewsbury Chronicle,* July 28, 1854.

11. Erasmus Darwin (*Zoonomia,* p. 134) expresses a similar idea: "In respect to free will, it is certain, that we cannot will to think of a new train of ideas, without previously thinking of the first link of it. . . ."

12. Probably Dr. John Ash, "respected Birmingham physician," and member of the Lunar Society with Dr. Erasmus Darwin (ref.: Schofield, Robert E., *The Lunar Society of Birmingham,* Clarendon, Oxford, 1963, pp., 39, 87, 88, 124, 227–228, and 323). Dr. John Ash obtained a position in Birmingham for a Dr. William Withering, with whom both Dr. Erasmus and Dr. Robert Darwin were to have bitter public disputes.

13. High Ercal, approximately 6½ miles northeast of Shrewsbury.

14. Ipecacuanha is mentioned in *Zoonomia,* p. 55, as affecting the sphincter of the anus.

15. The expression "Ira brevis furor est" (Rage is a brief insanity) occurs also on page 445 in "An Introduction to the Philosophy of Consciousness," Part 2, *Blackwood's Edinburgh Magazine,* 43:437–452 (Anonymous), 1838. See also n. 53 and n. 122.

16. Probably Market Drayton, 18 miles northeast of Shrewsbury, on the road to Maer. Ternhill, 3 miles southwest of Market Drayton.

17. Pet dog of the Wedgwoods. Also mentioned in Henrietta Litchfield: *Emma Darwin. A Century of Family Letters* (Cambridge: University Press, 1904), in a letter from Henry ("Harry") Wedgwood (1799–1885) to his mother, written May 24, 1827: "What brilliant evenings [Jos. and Allen] must be spending together, what a flow of soul! I pity even Squib when I think of it."

18. Wynne, see "Questions for Mr. Wynne," below.

19. Lord Berwick, probably of Berwick Hall, 2 miles northwest of Shrewsbury.

20. Yarwin's grandfather Erasmus ". . . stammered greatly, and it is surprising that this defect did not spoil his powers of conversation." Erasmus' eldest son, Charles (1758–1778), also stammered. See Krause, Ernst, *Erasmus Darwin, with a Preliminary Notice by Charles Darwin* (London: John Murray, 1879).

21. Emily Catherine, Darwin's sister (1810–1866).

22. Darwin drew a bracket of emphasis along this sentence in the left margin in his notebook.

23. Reynolds, Joshua, *Seven Discourses Delivered in the Royal Academy by the President,* Cadell, London, 1778, Discourse VII, pp. 322–323.

24. Susan Elizabeth (1803–1866), Darwin's sister.

25. Humboldt compares Mexican sculpture with that of Egypt, India, Greece and Italy. See Humboldt, Alexander, *Political Essay on the Kingdom of New Spain, etc.,* 2 vols., transl. by John Black, London, 1811. (In the Darwin Collection, Anderson Room, Cambridge University Library, is a copy signed, "Charles Darwin, Bueons Ayres.")

26. In *Zoonomia,* p. 145, Erasmus Darwin says, "A Grecian temple may give us the pleasurable idea of sublimity . . ." and ". . . when any object of vision is presented to us, which by its waving or spiral lines bears any similitude to the form of the female bosom, whether it be found in a landscape with soft gradations of rising and descending surface, or in the form of some antique vases . . . , we feel a general glow of delight." See also n. 49.

27. Metaphorically viewing plants as animals seems to have been a family tradition, for Erasmus Darwin in *Zoonomia,* p. 102, says: "the individuals of the vegetable world may be considered as inferior or less perfect animals; a tree is a

congeries of many living buds, and in this respect resembles the branches of coralline, which are a congeries of a multitude of animals." Emma Darwin in a letter to Lady Lyell, August 1860, said, "At present he [Charles] is treating Drosera just like a living creature, and I suppose he hopes to end in proving it to be an animal." (Litchfield, Vol. 2, 1915, p. 177.) And beside the sand walk at Down is a large beech tree, the "Elephant Tree," so-called by grandchildren of Charles (Raverat, Gwen, *Period Piece: A Cambridge Childhood*, Faber and Faber, London, 1960, pp. 157–158).

28. FitzRoy, Robert, captain of H.M.S. *Beagle* during Darwin's five-year voyage around the world.

29. See *Zoonomia*, p. 356: "Now as labour strengthens the muscles employed, and increases their bulk, it would seem that a few generations of labour or of indolence may in this respect change the form and temperament of the body."

30. Fox, William Darwin, Darwin's second cousin, fellow student at Christ's College, and intimate who introduced him to entomology.

31. Probably (Nathan) Hubbersty, Assistant Master at Shrewsbury School, 1826–1828.

32. Page crossed out with four diagonal lines, and the word "inaccurate" written across the page.

33. Burke, Edmund, *Philosophical Inquiry into the Origin of Our Ideas of the Sublime and Beautiful: with an Introductory Discourse Concerning Taste* (London, 1757), Part I, Section XIV, "The Effects of Sympathy in the Distresses of Others."

34. Nancy, Darwin's old nurse.

35. Frances Mackintosh (1800–1899) married her cousin Hensleigh Wedgwood.

36. Caroline Sarah (1800–1888), Darwin's sister. Nina, a pet dog. Caroline married her cousin, Josiah Wedgwood.

37. Clayton, Caroline Wedgwood's home, near Etruria.

38. Note similarity of this paragraph to the following from *Zoonomia*, p. 57: ". . . the whole skin is reddened by shame, and an universal trembling is produced by fear; and every muscle of the body is agitated in angry people by the desire of revenge."

39. Hensleigh Wedgwood, Darwin's brother-in-law.

40. ". . . children long before they can speak, or understand the language of their parents, may be frightened by an angry countenance, or soothed by smiles and blandishments." *Zoonomia*, p. 146.

41. Scott, Walter, *Guy Mannering or, the Astrologer,* 3rd ed., Ballantyne for Longman, etc., Edinburgh, 1815, Vol. 3, p. 26: "It is even so with me while I gaze upon that ruin, nor can I divest myself of the idea, that these massive towers and that dark gateway . . . not entirely strange to me. Can it be that they have been familiar to me in infancy, and that I am to seek in their vicinity those friends of whom my childhood has exchanged for such severe taskmasters?"

42. "Remains of the Rev. Richard Hurrell Froude, M.A., Fellow of Oriel College, Oxford, 2-vols. 8vo. London: 1838." Art. X.–2. *The Edinburgh Review, or Critical Journal,* for April 1838 . . . , 67:525–535, July 1838, pp. 528–529.

43. Sykes, *op. cit.*

44. Westwood, J. O., "On the Earwig," *Transactions of the Entomological Society of London* 1 (Pt. 3) :157–163, 1836.

45. Newport, G., "On the Predaceous Habits of the Common Wasp, *Vespa vulgaris,* Linn.," *Transactions of the Entomological Society of London,* 1 (Pt. 3) :228–229, 1836.

46. *Zoonomia*, p. 183: "One circumstance I shall relate which fell under my own eye, and showed the power of reason in a wasp, as it is exercised among men. A wasp, on a gravel walk, had caught a fly nearly as large as himself; kneeling on the ground I observed him separate the tail and the head from the body part, to which the wings were attached. He then took the body part in his paws, and rose about two feet from the ground with it; but a gentle breeze wafting the wings of the fly turned him round in the air, and he settled again with his prey upon the gravel. I then distinctly observed him cut off with his mouth, first one of the wings, and then

the other, after which he flew away with it unmolested by the wind."

47. Lewis, R. H., "Case of Maternal Attendance on the Larva by an Insect of the Tribe of Terebrantia, Belonging to the Genus *Perga*, Observed at Hobarton, Tasmania," *Transactions of the Entomological Society of London*, 1 (Pt. 3) :232–234, 1836.

48. See Darwin, Charles, *Journal of Researches into the Geology and Natural History of the Various Countries Visited by H.M.S. Beagle, under the Command of Captain FitzRoy, R. N. from 1832 to 1836*, Colburn, London, 1839, pp. 435–436: "My geological examination of the country generally created a good deal of surprise amongst the Chilenos . . . [they] thought that all such inquiries were useless and impious; and that it was quite sufficient that God had thus made the mountains." As late as 1861 Darwin again used the same expression in a letter to Lyell: "It reminds me of a Spaniard whom I told I was trying to make out how the Cordillera was formed; and he answered me that it was useless, for 'God made them.'" Darwin, Francis, and Seward, 1903, *op. cit.*, Vol. 1, p. 192.

49. *Zoonomia*, p. 145: "Our perception of beauty consists in our recognition by the sense of vision of those objects, first, which have before inspired our love by the pleasure, which they have afforded to many of our senses; as to our sense of warmth, of touch, of smell, of taste, hunger and thirst; and secondly, which bear any analogy of form to such objects." And on page 253: "So universally does repetition contribute to our pleasure in the fine arts, that beauty itself has been defined by some writers to consist in a due combination of uniformity and variety." See also n. 26.

50. Martineau, Harriet, *How to Observe. Morals and Manners*, Knight, London, 1838, p. 22: "A person who takes for granted that there is a universal Moral Sense among men, as unchanging as he who bestowed it, cannot reasonably explain how it was that those men were once esteemed the most virtuous who killed the most enemies in battle, while now it is considered far more noble to save life than to destroy it." And on page 23: ". . . every man's feelings of right and wrong, instead of being born with him, grow up in him from the influences to which he is subjected."

51. Mackintosh, James, *Dissertation on the Progress of Ethical Philosophy Chiefly During the Seventeenth and Eighteenth Centuries*. Edinburgh, 1830, "There is no tribe so rude as to be without a faint perception of the difference between right and wrong. There is no subject on which men of all ages and nations coincide in so many points as in the general rules of conduct. . . .

52. Martineau, *op. cit.*, p. 27: "The traveller having satisfied himself that there are some universal feelings about right and wrong, and that in consequence some parts of human conduct are guided by general rules, must next give his attention to modes of conduct, which seem to him good or bad, prevalent in a nation, or district, or society of smaller limits. His first general principle is, that the law of nature is the only one by which mankind at large can be judged. His second must be, that every prevalent virture or vice is the result of the particular circumstances amidst which the society exists."

53. Possibly "An Introduction to the Philosophy of Consciousness," *Blackwood's Edinburgh Magazine*, 43 (Pt. 1) :187–201; (Pt. 2) :437–452; (Pt. 3) :784–791, 1838, p. 199: "It is reserved for man to live this *double* life. To exist, and to be *conscious* of existence; to be rational, and to *know* that he is so." See also n. 15.

54. Dr. Dewar; see Abercrombie, John, *Inquiries Concerning the Intellectual Powers and the Investigation of Truth*, 8th ed., Murray, London, 1838, pp. 301–302, for discussion of ignorant servant girl, mentioned by Dr. Dewar, who during paroxysms showed surprising knowledge of geography and astronomy, a case of double consciousness.

55. Probably Benjamin Bynoe, "Assistant and Later Surgeon on H.M.S. Beagle." Barlow, Nora, ed., *Charles Darwin's Diary of the Voyage of H.M.S. Beagle*, Cambridge, University Press, 1933.

56. Probably John Allen Wedgwood (1796–1882), son of John and Louisa Jane Wedgwood.

57. Babington, George Gisborne, Esq., John Allen H. Wedgwood, and Charles Darwin are listed as members of the Athenaeum in 1838: *Rules and Regulations*

for the Government of the Athenaeum with an Alphabetical List of the Members, etc., London, III, 1834–1839. Perhaps Darwin has reference to Prof. Charles C. Babington, botanist.

58. Boz, i.e., Charles Dickens.

59. According to Dr. Sydney Smith (personal communication), in the early 1830s the FitzWilliam Pictures hung in the Free School (Perse) Hall, Cambridge.

60. Probably Sarah Elizabeth Wedgwood (1778–1856), Darwin's mother's sister, and sister of Josiah Wedgwood of Maer, the father of Emma, his cousin, whom he married five months later.

61. In the Transmutation Notebook E, p. 89, on Jan. 6, 1839, Darwin wrote, "Handwriting is determined by most complicated circumstances, as shown by difficulty in forging. Yet handwriting said to be hereditary, shows well what minute details of structure [are] hereditary." The upper half of pages 83–84 were excised by Darwin, but are restored in this text.

62. A circle drawn here.

63. This particular reference not positively traced, but see *The Canterbury Tales and Faerie Queene: With Other Poems of Chaucer and Spenser,* William P. Nimmo, ed., Ballantyne, Edinburgh, 1870, Book I, Canto IV, verse 33, p. 326: "As ashes pale of hue, and seeming dead, And in his dagger still his hand he held, Trembling through hasty rage, when cholor in him swell'd."

64. The Galignani edition has not been traced, but see *The Poetical Works of S. T. Coleridge,* 3 vols., Pickering, London, 1836, Vol. 2, p. 323, in "Zapolya, a Christmas Tale," Part II, Act IV, Scene 1 (Casimir's 13th speech).

65. Burke, *op. cit.,* n. 33.

66. Montaigne: this reference has not been traced.

67. Probably Woolwich, London (S.E. 18).

68. See Darwin, Charles, "Observations on the Parallel Roads of Glen Roy, and of Other Parts of Lochaber in Scotland, with an Attempt to Prove that They Are of Marine Origin," *Philosophical Transactions of the Royal Society of London,* Part 1:39–82, 1839, p. 73. Darwin has reference to the mental exertion expended in mentally reconstructing the series of events occurring over a long span of time in producing the coves indenting the otherwise regular parallel roads of Glen Roy. He had to struggle with such problems as: Did the land rise? Or did the sea subside? Or both? Was it due to marine or fresh-water action? Or glacial? Or a combination? And when and to what extent had erosion altered the original condition? See also: Barrett, Paul H., "Darwin's 'Gigantic Blunder,'" *op. cit.*

69. Two vertical pencil lines in margin beside this sentence.

70. Probably infant (Ernest Hensleigh, 1838–1898) of Hensleigh and Frances Wedgwood.

71. Probably William Lonsdale (1794–1871), Assistant Secretary and Curator of the Geological Society, 1829–1842. But possibly Rev. J. Lonsdale, member of the Athenaeum. Athenaeum, *Rules and Regulations,* 1834–1839.

72. Probably George Robert Waterhouse (1810–1888), Keeper of the Department of Geology in the British Museum, 1851–1880. Waterhouse published descriptions of numerous species of mammals and insects collected by Darwin during the voyage of the *Beagle.*

73. Possibly Rev. R. Jones mentioned *M* 155 and *D* 41.

74. Probably Alexander d'Arblay (1794–1837), son of Frances (Burney) d'Arblay; tenth wrangler in 1818 and Tancred studentship at Christ's College, Cambridge; deacon, 1818; priest, 1819; minister of Ely Chapel, 1836. See Stephen, Leslie, ed., *Dictionary of National Biography,* Smith, Elder, London, 1885, Vol. 2, p. 57.

75. Undoubtedly Rev. George Peacock (1791–1858), Lowndean Professor of Astronomy at Cambridge, Dean of Ely, member of the Athenaeum, and consulted as to appointment of naturalist for the *Beagle* prior to Darwin. See Barlow, *op. cit.;* and Athenaeum, *Rules and Regulations,* 1834–1839.

76. According to Dr. Sydney Smith (personal communication), added by Francis Darwin.

77. This parenthetical phrase inserted in pencil. The word "Horse" is difficult to decipher, and three or four words following it are illegible.

78. "CD 27" is written in following this line, in pencil, by Francis Darwin, according to Dr. Sydney Smith.

79. There are 3 vertical lines in the margin beside this paragraph in the notebook.

80. Mayo, Herbert, *The Philosophy of Living*, 2nd ed., Parker, London, 1838, p. 139, discusses when dreaming occurs during sleep, and disagrees with Brougham's conclusion that it is most frequent at the " 'instant of transition into and out of sleep.' "

81. *Ibid.*, p. 141: "Sir John Franklin remarks, when his party was in the extremity of physical exhaustion and physical suffering,—'Although the sensation of hunger was no longer felt by any of us;—yet we were scarcely able to converse on any other subject than the pleasure of eating.' " See also Franklin, John, *Narrative of a Journey to the Shores of the Polar Sea, in the Years 1819, 20, 21, and 22*, Murray, London, 1823, pp. 465–466: "The dreams which for the most part, but not always accompanied it [i.e., comfort of a few hours of sleep during prolonged starvation] were usually (though not invariably) of a pleasant character, being very often about the enjoyments of feasting." In Darwin's copy of *Zoonomia*, E. Darwin, 1794, now kept in the Anderson Room, Cambridge University Library, on page 23 is the following marginal notation penciled in in Darwin's handwriting: "This is strange as hungery men never dream of hunger." The text passage reads, ". . . in his dreams, he [a 50-year-old man who had been deaf for nearly 30 years] always imagined that people conversed with him by signs or writing, and never that he heard any one speak to him. From hence it appears, that with the [loss of] perception of sounds, he has also lost the ideas of them. . . ." Perhaps, when he wrote his margin note, Darwin had in mind, his statement on *M* 21: "People, my Father says, do not dream of what they think of *most* intently.—criminals before execution. —Widows not of their husbands—My father's test of sincerity."

82. Darwin apparently inadvertently continued *M* 110.

83. Darwin probably read in the Athenaeum Club library the copy of *The Philosophical Works of David Hume*, 4 vols., Constable, Edinburgh, 1825.

84. Smith, Adam, *Essays on Philosophical Subjects. To Which is Prefixed, An Account of the Life and Writings of the Author* by Dugald Stewart, Cadell and Davies, London, 1795, pp. xxvi–xxvii.

85. See Mayo, 1838, *op. cit.*, p. 145: ". . . I believe it [double consciousness] to exemplify sudden transitions to and from the state of somnabulism."

86. Both Herbert and Thomas Mayo are listed as members in 1838 in the Athenaeum. See Athenaeum *Rules and Regulations*.

87. Syms Covington, "Fidler and boy to Poop cabin" at the beginning of the *Beagle* voyage, and Darwin's servant from the second year of the journey until after their return to England.

88. Mayo, 1838, *op. cit.*, p. 140: "In dreams, that which most strikes us are their monstrous and capricious combinations, and our want of surprise at their improbability."

89. Abercrombie, *op. cit.*, pp. 296–298.

90. Browne, Thomas, *The Works of Sir Thomas Browne*, 6 vols., Geoffrey Keynes, ed., Faber and Gwyer, London, 1928. Vol. 1, *Religio Medici*, The First Part, Section 17, p. 23: "Surely there are in every man's life certain rubs, doublings, and wrenches, which pass a while under the effects of chance, but of the last, well examined, prove the meer hand of God."

91. Mayo, Herbert, *The Philosophy of Living*, Parker, London, 1837, p. 293: "Honesty is the recognition of the principle of property. It is remarkable that animals have this idea in its simplest form . . ."

92. Scott, Walter, *Memoirs of the Life of Sir Walter Scott*, Bart., J., G. Lockart, ed., 1838, Vol. 1, p. 127: "He [Will Clerk] to whom, as to all his family, art is a familiar attribute, wondered at me as a Newfoundland dog would at a greyhound which showed fear of the water."

93. Holland, Dr. Henry, afterward Sir Henry, second cousin to the Wedgwoods and Darwins, and physician to Queen Victoria.

94. See Darwin, R. W., *op. cit.*

95. N.B., page 128 in the notebook is dated September 4 and page 130 is dated September 3. Darwin apparently skipped a full page when he made the September 3 entry, then came back the next day to fill in the blank pages.

96. Lyell, Charles, *Principles of Geology, Being an Inquiry How Far the Former Changes of the Earth's Surface Are Referable to Causes Now in Operation*, 5th ed., 4 vols., Murray, London, 1837, Vol. 2, p. 418: "Some favourite dainty is shown to them, in the hope of acquiring which the work is done; and so perfectly does the nature of the contract appear to be understood, that the breach of it, on the part of the master, is often attended with danger."

97. Probably Richard Drinkwater, woolstapler of Frankwell, Shrewsbury, and Mayor of Shrewsbury, 1834–1835. Ref.: *Shropshire Archeological Society Transactions*, 4th. ser., vol. 9, 1923/4. (Thomas, A. L., Reference Librarian, Shrewsbury Public Library, Shrewsbury, England. Personal communication.)

98. Scott, *op. cit.*, Vol. 7, pp. 35–36: "May 28 [From diary]—Another day of uninterrupted study; two such would finish the work with a murrain. What shall I have to think of when I lie down at night and awake in the morning? What will be my plague and my pastime—my curse and my blessing—as ideas come and the pulse rises, or as they flag and something like a snow-haze covers my whole imagination? I have my Highland Tales—and then—never mind—sufficient for the day is the evil thereof."

99. Daubeny, Dr. [Charles G.], "On the Geology and Thermal Springs of North America," *The Athenaeum*, 1838, p. 652.

100. See Mackintosh, James, *Dissertation on the Progress of Ethical Philosophy. Chiefly during the Seventeenth and Eighteenth Centuries*. With a Preface by W. Whewell, 2nd ed., Black, Edinburgh, 1837, p. 280: "[Paley's] chapter [in *Moral and Political Philosophy*] on what he calls the *Law of Honour* is unjust . . . because it supposes honour to *allow* what it does not forbid. . . . He considers it 'a system of rules constructed by people of fashion. . . .'" In *The Descent of Man*, Murray, London, 1871, Vol. 1, p. 99, Darwin discusses at some length the Law of Honour. In general, it is conformance to expectations of one's peers.

101. Mitchell, Thomas Livingstone, *Three Expeditions into the Interior of Eastern Australia; with Descriptions of the Recently Explored Region of Australia Felix, and of the Present Colony of New South Wales*, 2nd. ed., 2 vols., Boone, London, 1839, Vol. 1, p. 295.

102. York Minster was one of three Fuegians brought back to Tierra del Fuego by Capt. FitzRoy and the *Beagle*.

103. Comte, Auguste, *Cours de Philosophie Positive*, 2 tom., 8vo. Paris: 1830–1835. [Review] *Edinburgh Review, or Critical Journal*, 67:271–308, 1838, p. 280: "'. . . each branch of knowledge, passes successively through three different theoretical states—the theological or fictitious state, the metaphysical or abstract state, and the scientific or positive state. . . .'"

104. Added in blue crayon between the lines.

105. Abercrombie, *op. cit.*, pp. 178–179: "The process of mind which we call reason or judgment, therefore, seems to be essentially the same, whether it be applied to the investigation of truth or the affairs of common life. In both cases, it consists in comparing and weighing facts, considerations, and motives, and deducing from them conclusions, both as principles of belief, and rules of conduct." Darwin's marginal notation: "Perhaps mathematical reasoning does not.—each step then does not require the memory and knowledge of all contingencies,—it is merely to find the step, & then to pursue this deep train." Abercrombie continued: "In doing so, a man of sound judgment proceeds with caution, and with due consideration of all the facts which he ought to take into the inquiry." Darwin's notation: "requires properly arranged memory."

106. Possibly Kenyon, John, *Poems: For the Most Part Occasional*, Moxon, London, 1838, p. 61: "Due honour to the stout-built Man of Prose!/Reasoner on facts! Who scorns to feel, but knows!/ Yet it be mine, who love not less the true,/ To lead, well feigning bards! my hours with you;/And sick, long since, of facts that

falsify,/And reasonings, that logically lie,/With you live o'er my wisely-credulous youth,/and in your fictions find life's only truth."

107. Martineau, *op. cit.*, p. 213.

108. Bynoe, *op. cit.*, n. 55. Perhaps Darwin has reference to the story of Fuegians eating their old women during famines. See *Charles Darwin's Diary of the Voyage of the Beagle*, edited by Nora Barlow, Cambridge University Press, 1933.

109. Hindmarsh, L., "On the Wild Cattle of Chillingham Park, *Annals of Natural History; or Magazine of Zoology, Botany, and Geology*, Jardine, Selby, etc., London, 2:274–284, 1839, p. 280: ". . . when any one [of the cattle] happens to be wounded or has become weak and feeble through age or sickness, the rest of the herd set upon it and gore it to death. This characteristic is an additional and strong proof of their native wildness." See also Darwin, Charles, *The Variation of Animals and Plants under Domestication*, London: John Murray, 1868; and Whitehead, G. Kenneth, *The Ancient White Cattle of Britain and Their Descendants*, Faber, Faber, London, 1953, Chapter IV, "The Wild Cattle of Chillingham Park."

110. Darwin means, "Vide *Notebook D*, p. 111." ("Case of association very disagreeable hearing maid servant cleaning door outside, as often as she touched handle, though really fully aware she was not coming in,—could not help being perfectly disturbed, referred to Book M.")

111. Alexander Monro, Darwin's lecturer in anatomy, University of Edinburgh.

112. See *Zoonomia*, p. 39.

113. Lavater, John Casper, *Essays on Physiognomy; for the Promotion of the Knowledge and the Love of Mankind.* Transl. by Thomas Holcroft, 2nd ed., *To which are added, One Hundred Physiognomonical Rules. A Posthumous Work by Mr. Lavater, and Memoirs of the Life of the Author* . . . by G. Gessner, 3 vols. (Vol. 3 in 2 parts), Symonds, Whittingham, London, 1804, Vol. 1, p. 86. "We ought never to forget that the very purport of outward expression is to teach what passes in the mind, and that to deprive man of this source of knowledge were to reduce him to utter ignorance; that every man is born with a certain portion of physiognomonical sensation, as certainly as that every man, who is not deformed, is born with two eyes; that all men, in their intercourse with each other, form physiognomonical decisions, according as their judgment is more or less clear . . ."

114. A single vertical line in blue crayon drawn down the margin of the page.

115. Added in blue crayon.

116. *Zoonomia*, p. 152: ". . . the horse, as he fights by striking with his hinder feet, turns his heels to his foe, and bends back his ears, to listen out the place of his adversary, that the threatened blow may not be ineffectual." In his personal copy beside this statement, Darwin wrote, "Sir C. Bell says because he looks back."

117. See *Zoonomia*, p. 430, for a discussion of antagonistic muscles.

118. See *Zoonomia*, p. 153: ". . . when we hear the smallest sound, that we cannot immediately account for, our fears are alarmed, we suspend our steps, hold every muscle still, open our mouths a little, erect our ears, and listen to gain further information. . . ."

119. Martineau, *op. cit.*, n. 50 and n. 107.

120. This reference not positively traced, but see Mackintosh, 1830, *op. cit.*, "Dr. Paley* (*Principle of Moral and Political Philosophy*) represents the principle of a moral sense as being opposed to that of utility. . . . Man may be so constituted as instantaneously to approve certain actions without any reference to their consequences; and yet reason may nevertheless discover that a tendency to produce general happiness is the essential characteristic of such actions." Darwin had met Sir J. Mackintosh in 1827 at Maer and each on that occasion developed great mutual admiration for the other. See *Autobiography*, p. 55. See also Mackintosh, 1837, *op. cit.*, p. 262: "The words *Duty* and *Virtue*, and the word *Ought*, which most perfectly denotes *Duty*, but is also connected with *Virtue* . . . become the fit language of the acquired, perhaps but universally and necessarily acquired faculty of conscience."

121. See *Zoonomia*, p. 148: ". . . the passion of fear produces a pale and cold skin, with tremblings, quick respiration, and an evacuation of the bladder. . . ."

122. "Consciousness," *op. cit.*, n. 15 and n. 53.

123. Watson, Hewitt Cottrell, published several papers on the geographical distribution of plants. See *Royal Society of London, Catalogue of Scientific Papers* (1800–1863), Eyre and Spottiswoode, London, Vol. 6, 1872, p. 280.

124. Couteur, Col. J. le (Sir John), *On the Varieties, Properties and Classification of Wheat*, Payn, Jersey, 1837 (Wright, London, reissue, 1838).

125. Jones, R., *op. cit.*, n. 73.

126. Hume, *op. cit.*, Vol. 4, "An Inquiry Concerning the Human Understanding."

127. Smith, Adam, *op. cit.*, n. 84.

M Notebook

Commentary by Howard E. Gruber

"Now if memory of a tune and words can thus lie dormant, during a whole life time, quite unconsciously of it,
surely memory from one generation to another, also
without consciousness, as instincts are, is not so very wonderful."* (*M* 7)

pp. 1–12   *Memory.* These opening pages are a set of notes on memory and
psychopathology. From the context, and from what follows, we
know that he is interested in exploring the following line of argument: The hereditary transmission of mental and behavioral
characteristics is analogous to memory; ideas, habits, and experiences which are stored in the brain in the form of long-enduring
memory traces must produce structural changes in the brain, which
are in fact those traces; such structural changes can be inherited,
and therefore constitute adaptation transmissible from one generation to another.

The collection of cases of long-dormant memories naturally
leads Darwin to emphasize experiences of early childhood, since
these enhance the point that memory traces are long-enduring
structural changes.

Darwin is interested in distinguishing between such memory
traces and conscious memories; it is the structural change which is
important, not the individual's consciousness of it. In this connection he exploits the medical materials his father makes available to him, dealing with memory in cases of mental illness and
senility. In such cases, the separation of conscious memory of the

---

* Darwin uses *wonderful* to mean "surprising." In the modern sense of *wonder*,
he always felt that all of nature was awesome, fascinating, and worth wondering about.

past from the present experience of the individual is brought out more sharply than in normal functioning. In order to make the link with normal functioning, Darwin will, in the immediately following passage, discuss the continuity between normal and abnormal.

The distinction between conscious memories and memory traces is closely related to the point that behavior and mental activity can be understood as parts of natural cause-and-effect relationships without necessarily being under the person's voluntary control. This point was important to Darwin in separating his position from Lamarck's. The latter believed that "will" was an essential part of all adaptive processes. Darwin used the concept of will much more hesitantly. Sometimes he thought it had no explanatory value and was just another name for ignorance of the actual causes of behavior. At other times he used the concept of will in ways resembling Lamarck's usage. Both authors, when writing of will in lower organisms, may have used the term only as a shorthand for the idea that an organism changes mainly through its own activity.

"My F[ather] says there is perfect gradation between sound people and insane.—that everybody is insane at some time." (*M* 13)

pp. 13–
22, with
an in-
trusion
from
p. 26

Psychopathology. Throughout these notebooks Darwin is concerned with the question: How much continuity can be found among the diverse forms of living things? In this passage he applies this general question to the relation between normal and abnormal psychological processes. Drawing upon his father's medical knowledge, and on the medical writings of his grandfather, he notes several distinct varieties of mental abnormality: insanity, delirium, senility, mania, intoxication. These have different causes and different symptoms.

Again, he introduces a note critical of rationalism: both normal and insane people do unreasonable things while *aware* that they are being irrational.

Even irrational acts and states have material causes, such as fearful events, drugs, and aging. His first explicitly stated "argument for materialism" is almost impossibly crude: "that cold water brings on suddenly in head a frame of mind analogous to those feelings which may be considered as truly spiritual." But more generally, Darwin's materialism is not at all crude: for instance, he considers a mental act to be a brain event, which can in turn serve as the material cause of some other event.

The attentive reader will find suggestions of many other themes in this sketchy passage—hints about repression, about the motor expression of emotions and thought, and about Darwin's own character. But the main thrust seems to be the emphasis on the *continuity* between sane and insane: "ira furor brevis est"—anger is a brief insanity.

> "One is tempted to believe phrenologists are right about habitual exercise of the mind, altering form of head, & thus these qualities become hereditary." (*M* 30)

pp. 23–32   *Hereditary, environment, and free will.* These pages and those that follow form a unit in which Darwin is exploring a number of related ideas about the inheritance of acquired mental characteristics. As his position takes shape, he draws on some material from the field of aesthetics, and then, in pp. 33–41, begins to examine that subject in greater detail.

Darwin has been discussing purely cognitive functions. Now he turns his attention to emotions—pleasure, pride, shame, aesthetic feelings. Dogs experience a variety of emotions, not unlike humans. In insanity, the disturbance of emotions ("affections") is a more prominent symptom than the failure of cognitive functions (memory). Emotions influence the bodily frame, and vice versa.

Many events in the life history produce changes that are transmissible through inheritance to the next generation. He mentions psychological traits, mutilations, and deformities. But his position is qualified in a number of ways. Only some deformities are heritable; some psychopathology may "remain many generations" in a family—but not forever. Even more important than these qualifications is an emerging idea about the way in which heredity works through interaction with environmental factors. Men and other animals are born with certain appetites, but these are non-specific ("appetites urge the man, but indefinitely . . ." *M* 31); there is room for adaptive changes to occur in the life history, and these changes may be passed on by one of several means.

Birds are born with a capacity to sing: those in England do so, but those in Tierra del Fuego do not. This shows that birds have "hereditary knowledge like that of man . . ." (*M* 32) The reader should notice that in choosing this example Darwin may be shifting away from his earlier certainty that man is unique in his capacity for cultural transmission of knowledge. (*M* 27)

Another example of the interaction of heredity and experience captures his attention. All men, he believes, have "an instinctive feeling" for beauty. What it is that particular men admire depends on particular circumstances which determine the "acquiring" of particular notions of beauty.

But if it is true, as he argues in this passage, that "every action [is] determined by hereditary constitution, example of others or teaching of others" (*M* 27), then the question arises: what need have we for the notion of free will? Which thought or action of a large number of possibilities will actually occur may depend not on free will but on natural law: "I verily believe free will & chance are synonymous.—Shake ten thousand grains of sand together & one will be uppermost,—so in thoughts one will rise according to law." (*M* 31)

"Mine is a bold theory, which attempts to explain, or asserts to be explicable every instinct in animals." (*D* 26)

pp. 28, 29, 33–41

Beauty and imagination. We see from Darwin's general attitude that he has taken the whole province of living things as his subject matter. The topic of instinct is included, and as we have just seen, in his view, a feeling for beauty is instinctive. This provides adequate justification for him to take time out to consider the psychology of aesthetics and imagination. We now know how Darwin eventually used these ideas in writing the *Descent of Man,* but if we consider only what he wrote up to this point it is not clear exactly how these notes fit into the argument he is developing at the moment.*

He expands on a remark made by his sister, leading to the following points: It is worth distinguishing among various related mental processes—memory, imagination, and invention. Early memories are largely of things seen, and in the form of visual images, "a set of sketches." Again, he makes a point about the distinction between what is stored in memory traces and what can be voluntarily remembered—children can recognize a familiar story that they "have not imagination enough to recall." (M 28)

The preceding points are intrusions in another line of thought, with his remark on page 32 about beauty as an instinct providing the bridge between his just previous discussion of memory and instinct and the present discussion of beauty and imagination.

pp. 33–41

Music and poetry. The brief remark on the continuity from music to singing to poetry seems cryptic standing in isolation. The reader

---

* Written between July 23 and August 15, 1838. Pages 1–42 of the M notebook were written from July 15 to July 21, 1838.

ought to set it alongside the argument in the *Descent of Man,* written over thirty years later—linking the courtship sounds of lower animals, the emergence of music in man out of these primitive tones and rhythms, and the evolution of language and poetry out of these sexual beginnings. One ought not to read the whole of Darwin's later argument into this one paragraph; it may be enough to say that he is here sensitive to yet one more of nature's continuities.

In the next few paragraphs, "castles in the air" seems to mean speculative daydreaming; "invention" seems to mean disciplined thought. Either can be hard work; it is useful to expand one's horizons by letting "castles" precede "invention." But sane and productive people will keep the distinction clear. He appends a few sentences harking back to his earlier thoughts on the continuity between the sane and the insane—the distance is not very great between normal daydreaming and insanity.

He now writes an organized essay on the "pleasures of scenery." The main points are clear enough and need no re-statement. They express the same interactionist view of the relation between heredity and personal experience alluded to earlier, but here with specific reference to a theory of aesthetics. Some of the objective conditions conducive to aesthetic pleasure are "absolute" or "instinctively beautiful"—notably harmony of color, symmetry of form, and visual rhythms; other factors, "the pleasure of imagination," draw upon the life experience of the individual.

We ought not pass over one significant sentence in a comment on two of his sisters: "Granny says she never builds castles in the air—Catherine often, but not of an inventive class." (*M* 33) To which ought to be added, of course, his own implicit description of himself as one who can do both, and in an original and productive sequence. Thus we have at least three major types of individual. This is an improvement on his discussion of visual imagery (*M* 28–29), where he generalized too much from his own personal experience. We see in this thought another expression of Darwin's interest in variation among the members of one species.

In this rich passage, then, we find reflections of many of the major themes which must be woven together to make the arguments of the *Origin* and *Descent:* variation, continuity, and hereditary transmission. There is very little suggestive of adaptation—perhaps he has not yet seen clearly the connection between the sense of beauty and the reproductive process that figures so largely in his later thought. ". . . I have shewn that the brains of domestic rabbits are considerably reduced in bulk, in comparison with those of the wild rabbit or hare; and this may be attributed to their

Charles Darwin and his sister Catherine in 1816. *Courtesy of American Museum of Natural History.*

having been closely confined during many generations, so that they have exerted their intellect, instincts, senses and voluntary movements but little."* (*Descent*, p. 55)

pp. 42–57 Now Darwin's awareness of the materialist point of view which he has been developing rises sharply. He writes briefly of two cases in which physical events produce mental changes—one idiot was cured by eating white lead, another by a fall.

He makes three points which, taken together, would constitute

---

* Darwin reported these results of a series of measurements he had made in *Animals and Plants*, Vol. 1, pp. 124–129.

a "law of exercise" in which effort is necessary for producing mental change, and repeated effort transforms a voluntary act into an involuntary one. Thus, motions and thoughts become habitual; intentional recollections depend on associations, which are like habits, "probably a physical effect of brain." (*M* 46)

There is a long passage on emotions. They are instinctive. Animals have emotions. They are involuntary. In collecting a number of observations on the subject, Darwin arrives at a hypothesis astonishingly like the famous James-Lange theory of emotions, of much later vintage: the external event to which the person or animal responds normally evokes from him some overt action; this act in turn has direct consequences for the physical state of the organism; these become the concomitants of emotion even when the overt action is omitted: "The sensation of fear is accompanied by troubled beating of heart, sweat, trembling of muscles, are not these effects of violent running away, & must not running away have been usual effects of fear." (*M* 57)

As though in a sudden moment of self-recognition, having completed this passage and realizing the contrast between himself and his contemporaries, he draws a line across the page and below it writes, "To avoid stating how far, I believe, in Materialism . . ." (*M* 57)

> ". . . then thinking consists of sensation of images before your eyes, or ears (language mere means of exciting association) or of memory of such sensation, & memory is repetition of whatever takes place in brain, when sensation is perceived." [*M* 61–62]

pp. 58–62 Darwin's attention seems to play almost rhythmically, first on one subject, then on another, returning to previously considered topics in the new light of intervening thoughts. These pages represent a kind of review with a few quick new insights.

In two sentences he makes it fairly clear that his earlier remarks on the continuity between music and poetry do have a bearing on the evolution of language: "It is known that birds learn to sing & do not acquire it instinctively. May not this be connected with their power of acquiring language." (*M* 58)

But he cannot permit himself to go too far in the direction of emphasizing learning rather than the hereditary transmission of ideas; if he does, he will have to abandon the notion he is now entertaining that ideas survive in memory in the form of structural alterations of the brain, which can be passed on from one generation to another. He therefore adds a few examples to his collection of complex instincts: babies recognize facial expressions and birds recognize cats without prior experience of them; walking in animals is instinctive.

Having touched thus briefly on instinct and learning in their relation to materialism, he returns to the subject of emotion, and describes two cases of overcompensation. In one of them, a revealing remark about himself, he worries a little about the corrosive effects of his own ingratiating behavior: "In making too much profession . . . of gratitude . . . I was tending to make myself in *act* less grateful." (*M* 60) This remark is embedded in some musings about unconscious feelings and seems to reflect an insight he has had about himself.

The highly imagistic theory of thinking expressed on pages 61–62 goes further than before simply by being more succinct. It becomes clearer than ever that for him the production of thought through the transformation of perceptions into images and memories is a function of a material organ of the body, the brain.

> "—good Heavens is it disputed that a wasp has this much intellect . . ." [*M* 63]

pp. 62–64 (65–68, excised, not found)
Instinct versus intellect. He has been reading entomology, with special interest in discussions of seemingly intelligent behavior in insects. In trying to sort out the distinction between instinctive and intelligent behavior, he argues that ants had not been behaving instinctively when they met a new situation by leaping across a gap: "if ants had at once made this leap it would have been instinctive, seeing that time is lost & endeavours made must be experience & intellect." (*M* 63)

Against a claim that a certain peculiarity of behavior in wasps cannot be intelligent because mistakes are made, Darwin argues from his own experience: *he* makes mistakes by persisting in an intelligent act which has become a habit, and *he* is intelligent—why not wasps?

From a letter to Charles Lyell, written August 9, 1838, one gathers that the *Transactions of the Entomological Society* were borrowed somewhere by Darwin for Lyell, to whom he forwards the work. During these days he has been reading Lyell's new *Elements of Geology* very thoroughly, and taking pleasure in the fact that his own *Journal of Researches,* as yet unpublished, has been liberally cited by Lyell. (*LL* 1, 295)

In *The Descent of Man* Darwin discusses the role of very primi-

M Notebook, p. 74. *Courtesy of Cambridge University Library.*

tive parental behavior in the evolution of morality. He uses the example of the earwig in both the earlier and the later comments on "blind storge," i.e. instinctive parental affection for the young. (*Descent,* 106)

> "The above views would make a man a predestinarian
> of a new kind, because he would tend to be an atheist.
> Man thus believing, would more earnestly pray . . .
> would be most humble . . . would strive to improve.
> . . ." [*M* 74]

pp. 69–75 *God's will, man's will, and chance.* Tantalizing, what we have of this passage begins mid-sentence. Darwin seems to have been reading a review of Comte's *Positive Philosophy,* the publication of which was begun in 1830. He sketches briefly Comte's notion that each branch of science passes through three stages—theological, metaphysical, and lawful. What really catches his attention is the first stage. He finds it "remarkable that the fixed laws of nature should be universally thought to be the *will* of a superior being." (*M* 69) From this anthropological fact, others may conclude that such a being exists; for Darwin it leads elsewhere: "one suspects that our will may arise from as fixed laws of organization." (*M* 70) In other words—man experiences his own actions as caused by his own will; personifying nature, he ascribes the same sort of will to a higher being whose will is expressed in the laws of nature; the very regularity of this psychological process suggests that it is caused by man's biological structure.

He goes on from this thought to make a series of other points about human will: Its power is limited—we cannot consciously control the expression of emotion. From observations of animal behavior, one "cannot doubt" that a puppy has free will in the same sense as a person; if a puppy has free will, why not all other animals, even an oyster, and maybe even plants? But Darwin's conception of free will does not express some lurking religious sentiment; on the contrary, he *denies* the necessity to invoke a superior being in order to account for it. And he goes on to say that the atheism toward which he is tending would not be harmful, would in fact make men morally better.

Interwoven with these thoughts are remarks about habit and instinct. The main burden seems to be the limitation of reason and will—habits change slowly—and the derivative character of will: out of the changes in physical organization of the individual, resulting from changes in habit, arise both hereditary transmission of habits and those changed mental characteristics which we experience as our "will."

There may be some connection between this passage and the one which follows. He has drawn a line separating them. But one deals with Auguste Comte and the next with Harriet Martineau. She was interested in Comte, and eventually translated his work into English. Darwin's brother, Erasmus, was a good friend of Harriet Martineau's, and Charles knew her too; they were all in the Carlyle circle.

It is striking to discover the identical grouping of ideas repeated in a letter from Darwin to Lyell in 1861. He is complaining that Asa Gray and Sir John Herschel, and perhaps Lyell too, cling to the idea of providential intervention in the natural order. In language and example not very different from the M notebook, he reminds Lyell that the Chilenos whom Darwin met during the *Beagle* voyage thought that God had made the Cordilleras. Darwin knows this will reach Lyell, the geologist most responsible for the uniformitarian view that the mountains were formed by the operation of the fixed laws of nature. He adds: "I must think that such views of Asa Gray and Herschel merely show that the subject in their minds is in Comte's theological stage of science." (*ML* 1, 192)

From this point on, throughout the remainder of the M notebook, the content grows increasingly psychological in character, and often remote from the main arguments which later emerge in the *Descent of Man* and in the *Expression of Emotions*. There is a general connection, of course: if man and mind are part of the entire network of evolutionary process, a scientific psychology based on materialistic premises must be possible. What better way for Darwin to test this proposition than to make his own efforts to construct such a science? He seems to sense, although it is nowhere explicitly stated, that psychology may entail special problems. Taking time out to look at these in their own right takes him pretty far from the main line of his evolutionary thought. These notes contain fascinating sketches of possible psychological treatises—on the causes of happiness, on dreams, on unconscious processes, on imagery, association, and creative thinking.

In the two later books, *Descent* and *Expression*, these topics are just barely alluded to. If we want to see the potential psychologist in Darwin, it is in these notes that we must look.

Naturally, Darwin reiterates his earlier arguments concerning the transition from habit to hereditary change in structure and instinct. To simplify the analysis, and to avoid needless repetition, I will from now on take it for granted that this argument is part of Darwin's thought and pass over it quickly unless a noticeable change occurs.

"The possibility of two quite separate trains going on in the mind as in double consciousness may really explain what habit is." [M 83]

pp. 75–83

Mind and body. Darwin has been reading Harriet Martineau and reflects on a statement of hers espousing an extreme of moral relativism. In his view, all men have some moral sense, although the particulars may vary from one group to another. Most important of all, the evolution of conscience is a "natural consequence" of man's being a social animal. Conscience is not something infused in man by a higher being. Morality evolves because it has survival value.

He turns his attention to an article on consciousness. He is intrigued with the non-rational—insanity, drunkenness, unconscious processes, double consciousness (which seems to include what we would now call "split personality" as well as more restricted phenomena such as automatic writing). In half a sentence he sketches a nice example of projection—a man he knew at Cambridge who, when he was in his cups, accused others of calling him a bastard, and actually was one.

What do these fragmentary anecdotes of the irrational add up to? Darwin seems to be saying that thought is the activity of the brain. The brain is a material organ. Therefore it is subject to a variety of physical influences. It is not made up of some mystical mind-stuff that acts instantaneously as a coherent whole; being material, its parts can act somewhat disjointedly.* Even an everyday habitual action is a case in point: one part of the brain controls certain actions of which another part is unaware, so that habits can run off smoothly without disrupting another train of thought or action.

A headache, its cause and cure, points up Darwin's ideas about the relation between mind and body: hard reading (about Comte, the turgid philosopher) gave him a headache; easy reading (Dickens) cured it. In another incident the very same day, the smell of an oil painting in London evoked an old memory of a museum in Cambridge. Here too we see his reliance on an empirical psychology in which physical stimuli are the cause of mental events, as well as a fleeting reference to his beloved idea of the permanence of mental changes.

Notice the ease with which he slips from the word *brain* to the word *mind* in the two similar sentences about double consciousness, written a few days apart. For him, now, the brain *is* the mind.

"—He who understands baboon would do more toward metaphysics than Locke." [*M* 84]

pp. 83–89

Emotion in animals and man. Darwin has been to the zoo. He has been there with his eyes open for animal expressions of emotion, linking man with the rest of creation in yet one more way. And he has found enough to make him exult, for it is he who will understand the baboon and outstrip the great empirical philosopher John Locke.

In this one paragraph he touches on a full range of emotions— joy, fear, and sexual desire. Not, at least in these notebooks, a sexually inhibited Victorian, he writes easily that the "smell of one's own pud" (pudenda, i.e., privy parts) is not disagreeable. He is at one with other male animals who smell their females' sexual parts.

The attentive reader will be puzzled at Darwin's paragraph on national character. He is probably not using the term in its modern social-psychological sense. Rather, it seems to refer to his

---

* It was not until 1850 that Hermann von Helmholtz first measured the speed of the nerve impulse. Before that time it was widely thought that the nerve impulse traveled at practically infinite speeds. If that were the case, the brain might well be a mysterious mind-stuff acting as a whole. Helmholtz was certainly aware of the philosophical implications of this work, since, as we have seen, he was a lifelong, explicitly committed materialist.

ideas about the geographical distribution of all natural fauna and flora. His views are hinted at in the books about the *Beagle* voyage, and expanded in the *Origin of Species,* where he concludes, "We see the full meaning of the wonderful fact, which must have struck every traveller, namely, that on the same continent, under the most diverse conditions, under heat and cold, on mountain and lowland, on desert and marshes, most of the inhabitants within each great class are plainly related; for they will generally be descendants of the same progenitors and early colonists." (*Origin,* 477). In the present paragraph we see that very early in his thinking, and in similar language, Darwin had applied this idea to man.

In the brief passage on sympathy he seems to accept Burke's somewhat misanthropic view of the source of that emotion. In the *Descent of Man* he treats the subject quite differently, as a more straightforward sharing of sorrow and concern, having definite survival value, "for those communities, which included the greatest number of the most sympathetic members, would flourish best, and rear the greatest number of offspring." (*Descent,* 107)

"effort consists in keeping one idea before your mind steadily . . . analogous to muscle in one position, great fatigue." [*M* 90–91]

pp. 89–92   *Thought and effort.* Thinking is hard work, Darwin has found. In the manner of a good scientific psychologist, he rules out various hypotheses as to the cause of the difficulty: It is not hard to think intently. It is effortless to attend to something present. If, as he believes, the difficulty lies in restricting yourself to one subject which is not present to the senses, then it ought to follow that even a simple subject, such as the idea of scarlet, ought to be difficult. This strikes him as worth an experiment, and he underlines "experimentize" with a double stroke.* I know of no such experiment by Darwin, but it would not surprise me to discover a scrap of paper, in his writing, describing it. He experimented with many things.

His father's only published contribution to science was a study of visual after-images.† From Darwin's imagistic theory of thought,

---

* Notice the analogy to Darwin's view, developed just a little later, that a given parcel of earth can support more inhabitants if they are of diverse species than if they are of one kind.
† R. W. Darwin, "On the Ocular Spectra of Light and Colours," *Philosophical Transactions,* Vol. 76, 1785, p. 313. Reprinted in *Zoonomia,* pp. 534–466. *op. cit.*

it should follow that abstract ideas are harder to think about than concrete objects. He sees immediately that this is not so: love and pain are as easy as scarlet. Another difficulty: lightning calculators must have strong imagery, but they are "not clever people," so the essence of inventive thought may not lie in imagery.*

"Seeing how ancient these expressions are, it is no wonder that they are so difficult to conceal." [*M* 93]

PP. 93–97 *Expression of emotions.* Although Darwin has touched several times on the subject of emotion, this is the first extended passage in which he concentrates on the *expression* of emotion, the motor activity involved in each expression, and the evolutionary continuity of man and other animals in the expression of emotion. He is now speaking in the same voice as in his book, published in 1872, *The Expression of Emotions in Man and Animals,* in which he wrote: "The community of certain expressions in distinct though allied species, as in the movements of the same facial muscles during laughter by man and by various monkeys, is rendered somewhat more intelligible, if we believe in their descent from a common progenitor." (*Expression,* 14)

In these notes he is making a series of characteristic points: Emotional expression is largely involuntary; attempts to conceal emotions usually come to naught. Animals have a wide range of emotional expressions similar to man. When he writes of a dog at play, he is implicitly saying that the significance of a particular expression can be understood only by looking at the behavior of the whole animal—a snarl means something special when the dog is playing.

While his main thrust is the search for continuity between man and animals, he raises at least one question about it—perhaps crying is "peculiar" to man. Years later, in *Expression,* this question surfaces again in the chapter "Special Expressions of Man: Suffering and Weeping," in which Darwin does what he must to concede the special character of man without conceding anything to the doctrine of special creation.

He has been watching a cousin's baby, newborn in 1838, and

---

* This passage foreshadows the pioneering work of Francis Galton on visual imagery; he too found a contradiction between his own persistent belief that imagery was very important in scientific thought and his empirical findings that it was not. Galton was Darwin's cousin, and for a time his collaborator. See F. Galton. *Inquiries into Human Faculty and Its Development* (London: Macmillan, 1883).

notices the very early appearance of certain emotional expressions and the absence of others. But the later appearance of sneering, presumably a vestigial snarl passed on to us by our evolutionary forebears, does not mean that it is learned; young animals have no canine teeth and consequently there is no need for them to bare them—snarling is reserved for animals old enough to hunt.

He sees an opportunity to bridge one of the largest gaps between man and other animals, language. Conceding the gap, he adds, "but do not overrate—animals communicate to each other," and goes on to list a number of cases of animal communication. In a quick synthesis of the two subjects of language and emotional expression, he writes, "they likewise must understand each other's expressions, sounds, & signal movements." (M 97)

"How does he account for dogs & men speaking in their sleep." [M 102]

pp. 99–
104, 110–
111

Dreams and thought. Darwin, reading several authors on the nature of dreams, enters briefly into the argument as to whether dreams occur only at the point of waking. From the telescopic, unreal character of dreams, he moves on to write about waking thoughts. Just as in the waking state fleeting series of thoughts are restful as compared with prolonged thought on one restricted subject, so dreams are a "rest of the mind."*

There are some new passages on double consciousness, or split personality, as usual coupled with anecdotes about the psychopathology of everyday life—two incidents of sudden, involuntary recall of long-forgotten events.

He makes a brief allusion to hybridization as affecting the hereditary character of the intellect: "half instincts when two varieties are crossed . . . shows that new instincts can originate." (M 101) This idea reflects a line of thought he is at this moment actively developing in the transmutation notebooks. He wants to find the causes of evolutionary novelty, and spends a good deal of effort trying to understand hybridization. In the present connection, he is satisfied to notice that thought "obeys same laws as other parts of structure." (M 101)

The extent to which Darwin at this time really believed in the

---

* This is not too far from modern findings about dreaming. Experiments in which special techniques are used to prevent the person from completing his dreams lead to marked hostility and disturbed behavior. Nor would the modern investigator find anything strange in Darwin's notion that dogs dream. Cats do too.

inheritance of mental characteristics acquired in experience is shown in what amounts to a eugenic proposal to improve the species not by selective breeding but by early education, since education received after the individual has done his share of procreation is "useless" in affecting future generations.

> "The ourang outang . . . threw itself down on its back
> & kicked & cryed like naughty child." [*M* 107]

pp. 105–109

The limits of the senses. Darwin is thinking about the psychological theories of the British empiricists. Something in Hume troubles him. Unconscious ideas are important in mental operations, according to Hume; but Darwin thinks that the things that interest us *least* are the ones most likely to produce unconscious ideas. A few pages further on he is troubled about Adam Smith's notion of sympathy. If we put ourselves in others' places we will share their suffering, and this will not give us pleasure. Hence the feeling of sympathy cannot easily be explained as deriving from the quest for pleasure.

Up to this point Darwin's psychological ideas have relied quite heavily on external influences on the individual—a blend of empiricism and simple hedonism. But now a different note is struck, if only in a tentative questioning way: "how little happiness depends on the senses." (*M* 109)

The zoo again. Observations on baboons, various monkeys. Now Darwin is paying attention to finer details of facial movements. And he gives a nice comparison of two species reacting quite differently to the same frustrating circumstance presented them by Darwin, both reactions recognizably similar to human emotions. It seems that on each of Darwin's visits to the zoo he used the occasion to do some informal experiment—hold up a mirror to a monkey, or make faces at an ourang, or in this case offer a nut and then withhold it.

> "parallel trains of thought necessary heirs of every action,
> & always running on in mind." [*M* 113]

pp. 111–117

Daydreams, dreams, and belief. A daydream (castle in the air, or just "castle") has slid over into real sleep and dreaming. Darwin tries to relate his own dreaming to what he has been reading on the subject: Like waking thought, dreams include memory and make some sense. The dream's vividness can be explained by the fact that it is a single train of ideas, whereas waking thought involves many parallel trains and the effort of relating them to each other.

A dream may be vivid, but it is not charged with credibility—it is neither believed nor doubted. Belief arises from awareness of the relationship among ideas and events, and this requires hard, active waking work. Darwin often expresses a similar view in his notes about the nature of scientific theory. For instance, about a week later, September 8, 1838, he writes, "In comparing my theory with any other, it should be observed not what comparative difficulties . . . [but] what comparative solutions & linking of facts." (D 71)

Once again, the importance of endogenous processes as compared with reactions to external influences is given a fleeting glance: "The senses are closed probably by sleep & not vice versa." (M 112) In other words, it is the internal state of the organism that determines its receptivity to outside stimulation, and not the cessation of such stimulation which produces the internal state.

This passage is a good one for noticing the way in which Darwin works back and forth between his own experiences, information gained in conversation with others, and whatever he is reading at the time.

". . . the delightful number of new views which have been coming in thickly and steadily,—on the classification and affinities and instincts of animals—bearing on the question of species. Note-book after note-book has been filled with facts which begin to group themselves *clearly* under sub-laws."*

pp. 117–
126

Happiness. Darwin is happy. He is happy at his work, which is the work of the mind, and this leads him to question a hedonistic view of happiness based on simple sensual pleasures: The pleasures of everyday life are so mild and so numerous that they dilute each other by their very number. Intense happiness depends on the accumulation of a number of pleasant experiences in memory, but ordinary pleasant experiences are not memorable and hence contribute little to the sum of intense happiness. The philosopher, Darwin himself, undergoes some pain, the pain of hard mental work, for which he is rewarded by the exquisite pleasure of the occasional deep insight.

---

* *LL* 1, 295–298. Charles Darwin to Charles Lyell, September 13, 1838.

Darwin goes on to remark that this view of the quest for truth as a source of pleasure departs from his Christian beliefs. It is one thing to justify intellectual effort on the grounds that truth brings you closer to God's natural order, that the pains of the search fulfill the command to mortify your flesh, and that the after-life is "almost the sole object" of this one. It is another matter, entirely, to lust for truth because the pursuit itself brings pleasure.

"Our descent, then is the origin of our evil passions!!— The Devil under form of Baboon is our grandfather!" [*M* 123]

pp. 117–126   *Passion*. Thoughts on the nature of happiness are interwoven with thoughts of "evil passion," such as revenge and anger.* Animals of lesser intelligence than man may have needed such motives for preservation of their kind, but man can afford to check them. Here Darwin is talking about the evolution of human ethics, and he finds it just as natural as evolution itself that ethics continue to evolve under "changing contingencies."

These thoughts are echoed in the *Descent of Man*. Darwin writes of an evolving social consciousness: "As man advances in civilization, and small tribes are united into larger communities, the simplest reason would tell each individual that he ought to extend his social instincts and sympathies to all the members of the same nation . . ." and eventually "to the men of all nations and races." (*Descent*, 122) In spite of some doubts, he argues for the occurrence of at least a partial hereditary "transmission of virtuous tendencies," making for the continuous improvement of man. (*Descent*, 124)

In the notebook passage under consideration, Darwin's thoughts move back and forth between matters of good and evil, but he does not speak explicitly of the struggle. In *Descent*, however, he writes: "As a struggle may sometimes be seen going on between the various instincts of the lower animals, it is not surprising that there should be a struggle in man between his social instincts, with their derived virtues, and his lower, though momentarily stronger impulses or desires." (*Descent*, 125)

---

* At the zoo, observing primates, Darwin is at least as interested in their sexual behavior as in their anger. One wonders which "evil passions" he had in mind. In the sentence quoted, beginning "Our descent . . .," he began by writing "Their" and then wrote the first-person pronoun over it, in a bold hand, like a statement of personal solidarity with sexual man. This was the last summer of his courtship.

> "Plato . . . says in Phaedo that our 'imaginary ideas'
> arise from the preexistence of the soul, are not derivable
> from experience.—read monkeys for preexistence."
> [*M* 128]

pp. 127–
131

These pages are a miscellany of brief observations and memo-
randa on association, animal intelligence, expression of emotions,
and imagery, and even one note on geology which Darwin probably transferred to another notebook. The scattered character of
these notes may be due to the fact that Darwin was in the throes
of finishing a difficult geological paper.*

The reference to Lyell, fleetingly disputing him, suggests an interesting distortion on Darwin's part. Lyell does not in fact find
the intelligence of elephants surprising. His *Principles of Geology*
contains a lengthy attack on the Lamarckian theory of the transmutation of species. In the passage cited in Darwin's notes, Lyell is
discussing apparent changes in animals wrought by man's efforts to
domesticate them. Lyell concludes, first, that the "Author of Creation" would naturally endow each species with some variability to
permit it to adapt to changing circumstances; and, secondly, that
the known variation, whether produced by domestication or any
other means, fall within very narrow limits and "imply no definite capacity of varying from the original type."† Darwin's argument with Lyell, therefore, is not about the abilities of elephants,
as his note suggests, but about the origin of species.

> "savages . . . consider the thunder & lightning the direct
> will of the God . . . & hence arises the *theological* age
> of science in every nation." [*M* 135]

pp. 132–
144

Origin of morality. Darwin entertains two distinct notions. On the
one hand, he argues that social behavior which has long been

---

* Finished on September 6, 1838 and published the following year: Charles
Darwin, "Observations on the Parallel Roads of Glen Roy and of other parts
of Lochaber in Scotland, with an attempt to prove that they are of marine
origin." *Philosophical Transactions of the Royal Society,* 1839, pp. 39–82.
Darwin maintained an abiding interest in this peculiar geological formation.
Eventually he had to admit that his explanation of its origin was incorrect.
See Paul H. Barrett, "Darwin's 'Gigantic Blunder,'" *Journal of Geological
Education,* Vol. 21, 1973, pp. 19–28. Among the Darwin *MSS* in Cambridge University is a small notebook written in light pencil and obviously used as a field
notebook for observations at Glen Roy in 1838. During this geological excursion
he continued to think about problems of evolution. Among the intrusions into
his geological train of thought is the remark, "The union of two instincts crossing most remarkable—?ever observed?—shows that brain makes thought . . ."
† Lyell, *Principles,* 1835, *op. cit.,* Vol. 2, p. 465.

valuable to a species becomes habitual and then instinctive. On the other hand, he seems to accept Comte's notion that the idea of God arises as a first explanatory principle when man confronts the unknown and the inexplicable. These ideas are not actually contradictory, since they concern two different domains, moral behavior and intellectual efforts to understand nature. The god whom Darwin is striving to depose is not so much the giver of moral law as the alternative to natural science.

Pages 133–34 are missing. The remnant of a sentence with which page 135 begins suggests that Darwin is still moving away from the more thoroughly empiricist philosophy of the earlier notes. External impressions are no longer enough; the powers of the mind must be considered; and out of all relation to experience we assign imaginary causes to events, to meet some need within us.

Tool-using in animals. Darwin has been to the zoo again, perhaps twice in a few days. He makes a few remarks on sexual curiosity in monkeys, and on the expression of emotions. But the point that seems to have caught his attention is tool-using and related acts in primates. A few pages back, after a previous visit to the zoo, he wrote, "I saw the ourang take up a stone & pound the earth." (*M* 129) Now he adds a number of other observations of animals using objects as implements. In *Descent,* the same facts are used to refute the argument that one of the unbridgeable gulfs between man and animals is man's use of tools.*

The brief note on his September 16 visit to the zoo is of interest because it shows that Darwin has passed from simply recording a few anecdotes about expression of emotions to classification of his observations and comparisons among related species.

Dream and belief. The fascinating contents of his dream of execution have been discussed above. To the modern reader it is almost incredible that Darwin pays little attention to these contents. Instead, he is still preoccupied with the question of the reality character of memories, images, and dreams. It would seem, however, that he was not entirely oblivious to the possible significance of the dream, for he has written in, probably at a later date, "What is the Philosophy of Shame & Blushing? Does Elephant know shame—dog knows triumph." (*M* 144)

---

* "I have seen a young ourang, when she thought she was going to be whipped, cover and protect herself with a blanket of straw." (*Descent,* 81) The 1838 note from which this is drawn is one of the excised passages (*M* 140), and in re-filing the slip of paper for use in writing *Descent,* Darwin has marked this incident in the margin with a double stroke in orange crayon.

> "The whole argument of expression more than any other point of structure takes its value from its connexion with mind (to show hiatus in mind not saltus [leap] between man & Brutes). . . . Hyaena pisses from fear so does man.—& so dog." [*M* 151–153]

pp. 144–156 and inside back cover

Laws of Expression. Darwin is now well into the subject of expression, continuing to collect observations, reading Lavater's lengthy treatise, and, most important of all, beginning to formulate the three principles of expression with which he later will begin his book *The Expression of the Emotions in Man and Animals.*

In *Expression* Darwin writes, "I. *The principle of serviceable associated habits.*—Certain complex actions are of direct or indirect service under certain states of the mind, in order to relieve or gratify certain sensations, desires, &c; and whenever the same state of mind is induced, however feebly, there is a tendency through the force of habit and association for the same movements to be performed, though they may not then be of the least use." (*Expression,* 28)

It is plain that Darwin's description of horses is intended as an illustration of this principle, along with numerous other examples he gives. His knowledge of horses, it should be remembered, was rather profound, as he had spent many months in the saddle on overland trips during the voyage of the *Beagle.*

Continuing, "II. *The principle of Antithesis.*—Certain states of the mind lead to certain habitual actions. . . . Now when a directly opposite state of mind is induced, there is a strong and involuntary tendency to the performance of movements of a directly opposite nature, though these are of no use; and such movements are in some cases highly expressive." (*Expression,* 28)

This principle is extremely important, indicating that Darwin's theory is not strictly utilitarian but accounts for much behavior on grounds of structure only indirectly related to utility. I am not sure Darwin was aware of the point, but the principle of antithesis is antagonistic to the argument from Design, and Darwin was always looking out for natural phenomena that would be imperfect or pointless from the point of view of an all-knowing Designer.

The third principle is something of a vague catch-all, referring to the "direct action of the nervous system," which is similar to the second in revealing the extent to which Darwin was free of utter reliance on desire and habit in his conception of emotion. There is an interesting shift in the organization of ideas from the

1838 notes to the 1872 volume. In the notes he writes as though all actions are governed *either* by will *or* by habit. In the book some actions are thought to be "due to the constitution of the Nervous System, independently from the first of the Will, and independently to a certain extent of Habit." (*Expression*, 29)

The same cat in two opposing moods. Illustrations from *The Expression of the Emotions in Man and Animals.*

p. 153   *Man and ape.* This brief passage is too elliptical for a clear interpretation. He appears to me to be saying that when others point out the "immense difference between man" and the higher apes, he will answer that, apart from language, the behavior of some extremely primitive human groups is not so different from that of the ourang-outang as to rule out the idea of continuity between man and animals.

p. 154   *Religion and science.* Darwin seems to be considering a possible psychological cause of the prevailing preference for the idea of a continually intervening Providence over the idea of a Creator "governing by laws." The latter view leads one toward an abstract attitude toward things as expressions of those general laws, and prevents the enjoyment of them as concrete objects to be admired for themselves, which is accomplished by comparing them "to the standard of our own minds." This passage is an interesting foreshadowing of his remark in later life that his own ability to experience aesthetic pleasures had diminished: "My mind seems to have become a kind of machine for grinding general laws out of large collections of facts. . . ." (*Autobiography*, 139)

It is noteworthy that in his *Journal* the entries for September and October 1838 do not mention the fact that toward the end of September he read Malthus' *Essay on Population* and from it got his first clear insight into the principle of natural selection. But he does write in the *Journal*, "All September read a good deal on many subjects: thought much upon religion. Beginning of October ditto."

p. 157   *Babies.* During the months in which he wrote the M notebook, Darwin was also occupied with many other things, among them courting Emma Wedgwood, whom he married on January 29, 1839. Eleven months later their first child, William Erasmus Darwin, was born. The baby immediately became the subject of a series of systematic observations, his father's pioneering effort in scientific child psychology. Darwin did not publish these observations until 1877. But inside the back cover of the M notebook, most likely written in 1838 before he married Emma, can be found scrawled a series of good questions, headed "Natural History of Babies."

N Notebook

What are sexual differences in monkeys.

Charles Darwin

PRIVATE

Metaphysics & Expression
Selected/for Species & Theory
Dec. 16, 1856

Looked through & all other Books May 1873

1   October 2ᵈ. 1838

Essays on Natural History

Waterton[128] describes pheasant springing from nest & leaving no tracks.— ((My Father says pea hens do [ditto].))—Wood pidgeons building near houses, yet so shy at all other times.— ((Birth Hill[129] shows it is *evergreens* they seek.)) Cock pheasant claps his wings *before?* crowing & only in breeding season & on the ground.—Cock fowl on the ground, at roost, in all seasons, & after? he has done crowing.[130]—instances of *expression*.—

Octob. 3ᵈ. Dog obeying instinct of running hare is stopped by

2   fleas, also by greater temptation as bitch: or dogs | defending companion (mem Cyanocephalus Sphynx howling when I struck the keeper) may be tempted to attack him from jealousy (Pincher & Nina) [131]—or to take away food etc etc. Now if dogs mind were so

N Notebook, p. 98. *Courtesy of Cambridge University Library.*

framed that he constantly compared his impressions, & wished he had done so & so for his interest, & found he disobeyed a wish which was part of his system, & constant, for a wish which was only short & might otherwise have been relieved, he would be sorry or have troubled conscience.—Therefore I say grant reason to | any animal with social & sexual instincts /& yet with passion/ he *must* have conscience—this is capital view.—Dogs conscience would not have been same with mans because original instincts different.—

3

Mem. Bee how different instinct[132] a solitary animal still different. Different nations having different moral sense, if it were proved [?] instead of militating against the existence of such an attribute would be rather favourable to it!!

Man moreover who *reasons* much on his actions, makes his conscience far more sensitive. ultimate effects of actions. | till at last he faces /instinct of/ hunger, of death & for the satisfaction of following conscience, obeying habits, & dread of misery of future thinking of injured moral sense.—

4

Notion of deity effect of reason acting on (<not social instinct>) but a *causation,* & perhaps on instinct of conscience, feeling in his heart those rules which he wills to give his child.—

Octob. 3ᵈ. Was told by W.[133] of Downing Coll. that he had seen chicken only hatched few hours placed on table & when fly ran past it, cocked its head, & picked it—Here then, that faculty, |

5

whether for position of axe of eyes, state of surface, or other means by which eyes, aided by experience is supposed in man to guide to knowledge, was transmitted *perfectly* to chicken so as to seize small moving object like fly.—young partridge can run even with its shell on back.—

To study Metaphysics, as they have always been studied appears to me to be like puzzling at astronomy without mechanics.—Experience shows the problem of the mind cannot be solved by attacking the citadel itself.—the mind is function of body.—we must bring some *stable* foundation to argue from.— |

6 October 4th. Seeing some drawings /in Lavater,[134] P. cii, Vol. III/ of excessively cross-half furious faces /which may be described as an exaggerated habitual sneer/ the manner in which whole skin or muscles are contracted between eyes & upper lip, is most clearly analogous to a panther I saw in garden uncovering its teeth to bite.—The *senseless* grin of passion is like the grin of the Hyaena from fear, no actual intention to bite at moment, but mere symbol of readiness, & therefore done in extreme.—

Looking at one's face whilst laughing in glass & then as one ceases, or stops the noise, the face clearly passes into smiles—laugh long prior to talking, hence one can help speaking, but laughing involuntary.— |

7 When one fear any bad news /[as] though in a letter/, why is person painted with mouth open.—why when person is listening is mouth open to hear well ((as one will perceive if in night trys to listen to growl of hounds)). as fear to /man as/ animals comes at distance, mouth is placed open.—Hence becomes instinctive to fear, as ears down to horse.—Horse snuffs the air /& snorts/, /& raises its head, & pricks its ears/ when afraid, though not every time really wishing to smell its enemy.—Man & dogs show triumph (& pride) same way walk erect & stiff, with head up.—

Why does suspicion look obliquely—who can analyse suspicion —yet who does not recognise look of suspicion, even child will do so.—

Contempt look obliquely so does dog when a little one attacks

8 him. | Contempt, when there is some anger /& respect to opponent/ is showed by same movement as sneering,—it is then more manner of *hurting* opponent by insulting his pride & is therefore of the snarling order.—But contempt mingled with disgust, when one opponent is considered as quite insignificant, & when pride makes person extremely self-sufficient,—the corner[s] of lower lip are depressed & *opposite* muscles used to when angry sneering is in progress—the hypothesis of opposite muscles will want much confirmation

9 A grave person closes those muscles, which wrinkle | when smile. —Hope is the expectant eye looking to distant object, brightened & moistened by emotion,—why does emotion make tears fall??

Lavater[135] says derision lies in wrinkles about the nose, & arrogance in upper lip.—Children having peculiar expressions is remarkable. the *pouting* & blubbring—*sulkiness* is same as pouting <but> lesser in degree, no smile, no frown showing thought, no compression of mouth showing action,—sulkiness all negative expression?

Expression of affection is accompanied by slight protrusion of lips, as if going to say "my dear," just what smile is to laugh.— |

10   I must be very cautious. Remember how Lavater[136] ran away with new [illegible]—Ye Gods!!—Says fleshy lips denote sensuality (p. 192, Vol. III, Octav. Edit.) —certainly neither a Minerva or Apollo would have them because not beautiful—is there anything in these absurd ideas—do they indicate mind & body retrograding to ancestral type of consciousness etc etc.—Lavater[137] (Holcroft Translat.) Vol. III, p. 37, quotes from Burke who says on mimicking expression of emotions, he has felt the passions of a face /& mind/ sympathetic with internal organs, as action of heart.

Malthus[138] on Pop. p. 32 origin of Chastity in women,—rationally explained—on the wish to support a wife a ruling motive. —Book IV, Chapt I[139] on passions of mankind, as being really

11   useful to them: this must | be studied before my view of origin of evil passions.—

Man getting sight slowly, but when in grown years, thinking he instinctively knows distances, is good instance of obtaining a faculty in the *form* of a true instinct, which is a *real* instinct in the chicken, just bursting from egg.—

Animals have necessary notions. which of them? & *curiosity* ((strongly shewn in the numerous artifices to take birds & beasts)) .—very necessary to explain origin of idea of deity.—Animals do not know they have these necessary notions any more than a Savage |

12   M. Le Comte's[140] idea of theological state of science. grand *idea:* as before having analogy to guide one to conclusion that any one fact was connected with law.—as soon as any enquiry commenced, for instance probably such a thing as thunder would be placed to the will of God.—Zoology itself is now purely theological.—

Origin of cause & effect being a necessary notion, is it connected

13   with <our> the willing of the | simplest animals, as hydra towards light being direct effect of some law.—have plants any notion of cause & effect /they have habitual action which depends on such confidence/ when does such notion commence?—

Children understand before they can talk, so do many animals. —analogy probably false, may lead to something.—

October 8th. Jenny[141] was amusing herself by getting out ears of corn with her teeth from the straw, & just like child not knowing what to do with them, came several times & opened my hand, & put them in—like child. Tommy's face, now ill, has *expression* of languor & suffering. |

14 The Cyanocephalus[142] when fondling the keeper, clasping /& rubbed/ his arm, & showed signs of affecting something like man.

Has the oyster necessary notion of space—plant though it moves doubtless has not.—

Turkey cock in passion & sends blood to its breast etc etc

All Science is reason acting /systematizing/ on principles, which even animals practically know ((art precedes science—art is experience & observation)) in balancing a body & an ass knows one side of triangle shorter than two. V. Whewell,[143] Induct. Science, Vol. I, p. 334. |

15 Does a negress blush—I am almost sure Fuegia Basket did ((& Jemmy when Chico plagued him)).[144] Animals I should think would not have any emotion like blush.—When extreme sensation of heat shows blood is pumped over whole body.—is it connected with surprise.—heart beginning to beat.—Children inherit it like instinct, preeminently so.—Who can analyse the sensation when meeting a stranger who one may like, dislike, or be indifferent about, yet feel shy.—not if quite stranger—or less so.— |

16 When learning facts for *induction,* one is obliged carefully to separate its memory from all ordinary lines of association, is totally distinct from learning *it by heart.*

Do not our necessary notions follow as consequences on *habitual* or instinctive assent to propositions which are the result of our senses, or our *experience.* ((Two sides of a triangle shorter than third; is this necessary notion; ass has it.—))

17 When one is /simply/ habituated /in life time/ to any line of action, or thought, one feels pain at not performing it (either if *prevented* or *overtempted*). ((Animals have shyness with strangers.)) ((As in case of temperance or real virtue, that is action which experience shows will be for general good, or in case of any fantastic custom.)) ((Probably bashfulness is connected with some disturbed habit.)) [Thus shepherd dog has pleasure in following its instinct & pain if held.—if tempted not to follow it, by greater temptation, if *memory of its own emotions* (which must be intimately united with *reason*) it would feel /subsequent/ sorrow, whatever the cause had been.][CD] Also when one is prevented performing *hereditary habit* (or moral sense, or instinct) one feels pain, & vice versa pleasure in performing it.— |

18 As soon as memory improved, direct effect of improving organization, comparison of sensations would first take place whether to

pursue immediate inclination or some future pleasure,—hence judgment, which is part of reason.

Octob. 19th. Did our language commence with singing—is this origin of our pleasure in music—do monkeys howl in harmony—frogs chirp in do [ditto]—union of birds voice & taste for singing with Mammalian structure. The taste for recurring sounds in Harmony common to to whole kingdom of nature. ((—American monkeys utter pleasant plaintive cry—)) |

19e   If I want some good passages against opposition of divines to progress of knowledge, see Lyell[145] on Scrope. Quarterly Review. 1827?

In Walter Scotts[146] life, Tom Purdie, (beginning of Vol. V) /illegible/ says "he knew no more what was pretty & what ugly than a cow"—((so it is with all uneducated.—)) Old man at Cambridge observed the ignorant merely looked at picture as works of imitation.—Hence pleasure in the beautiful (distinct from sexual beauty) is acquired taste.—Whilst music extremely primitive.—almost like tastes of mouth & smell.[147] |

20e   Understanding language seems simplest case of Association.—Elephant often given food & [when] word[s] open your mouth said, recognizes that sound as perfectly as a man.—Probably, language commenced in some necessary connexion between things & voice, as roaring for lion, etc etc. (in same way alphabet arose from letters, symbol of word beginning with the sound of letter) —crying yawning laughing being necessary sounds . . . not produced by will but by corporeal structure.—

Devotional feelings probably more distant power of the mind—
21e   superstition & charity & prayer, or *eloquent* request. | Reason in simplest form probably is single comparison by senses of any two objects—they by VIVID power of conception between one or two absent things.—reason probably mere consequence of vividness & multiplicity of things remembered & the associated pleasure as accompanying such memory.—

A Melody on flute & Epic poem, opposite ends of series or harmonious prose.— |

22e   Lutké[148] Voyage in Carolinas Vol II p. 132 offered to take a savage, said his wife would be grieved—"il leva les épaules et dit qu'il valait mieux rester a Farroïlap quelque mal qu'on y fût."—

Expression common to Savage & Frenchman, unaccompanied by dignity—"no mon dieu," with a shrug—"all I can say, I am very sorry so it is"—does not accompany *I will not*. I am sorry I *can-
23   not*.[149]— | without, however, very sincere grief,—"there is nothing more to be said."—"made no reply, but shrugged his shoulders & went away."—he implies negation, without violence, without as-

signing or understanding reason.—surprise with negation.—like shaking something off shoulder—or is it from *inspiration,* which accompanies surprise—& why does one inspire when surprise, can one resist blow better with body distended.—intolerable to be poked behind without ones chest being distended. touch a person

24 on the ribs & how he gulps in air.— | Again, a master says I will see you damned first." the man shrugs his shoulders & replies nothing. if he did go to reply, he would throw back his shoulder. he wishes to show, he is determined not to say anything. he presses his lips together & shrugs his shoulders & walks off. I think shrugging connected with many emotions.— (Explanation of sighing is probably correct, to relieve respiration when immensely immersed—mechanic apt to sigh—& hence carried on as trick) ((Shrugging aroused acting)) |

25 Octob. 25. Why is modesty, mixed with triumphant feeling so similar to shame after [being] asinine.—both accompanied by depending head, & active vessels of skin.—What difference is there between Squib[150] after having eaten meat on table, & criminal, who has stolen. neither, or both, may be said to have fear, but both have shame—Animals have not modesty. analyse this.—*Excellent*—my theory of blushing solves this.— |

26 The similarity of man's reasons shewn by similarity of the earliest arts. Mem. Stokes[151]—arrow heads etc etc

October 27th. Consult the VII discourse by Sir J. Reynolds.[152]— Is our idea of beauty, that which we have been most generally accustomed to:—analogous case to my idea of conscience.—deduction from this would be that a mountaineer born out of country yet would love mountains, & a negro, similarly treated would

27 think | negress beautiful.—[male glow worm doubtless admires female, showing no connection with male figure][CD]—As forms change, so must idea of beauty.—[Old Graecian living amongst naked figures, and observing powers common to savages???].[CD] The existence of taste in human mind is to me clear evidence, of the general ideas of our ancestors being impressed on us.—Surely we

28 have taste naturally all has | not been acquired by education. else why do some children acquire it soon. & why do all men ultimately?—

We acquire many notions unconsciously, without abstracting them & reasoning on them (as *justice??* as ancients did high forehead sign of exalted character???) Why may not our hereditary nature thus acquire some general notions, which are taste? |

29 Real taste in mouth, according to my theory must be acquired, by certain foods being habitual—& hence become hereditary; on same principle we know many tastes become acquired during life

time:—the latter correspond to fashions in ideal taste & the former to true taste.[153]—

Everything that is habitual, if hereditary, is pleasant.—Mental & Bodily.[154]

Consider case of grazing animals knowing poisonous herbs: & man not.—?no vegetable good /for man/ to eat poisonous? How did animals in Australia & America manage:—This shows doctrine (of [illegible]) has been carried too far. |

30   In all the foregoing cases most difficult to distinguish between prejudices of youth from *habit* & hereditary habits. & perhaps even latter may be vitiated, or rather altered.—

The Reason why new Buildings look ugly is because there is some connection between them, & great masses of rock.—I was much struck with this, when viewing Windsor Castle which rises

31   naturally & hence | sublimely from natural rise—I was also much struck in great avenue, resemblance to gloomy aisle of Churche.— These are Mayo's[155] ideas.—

In language, the possibility of poets describing gentle things in gentle language, & vice versa,—almost proves that at earliest times there must have been intimate connection between *sound* & language.—Chinese simplest language. Much pantomimic gesture?? Which would naturally happen.— |

32   Reynold's Works.[156] Vol. I, p. 226, "The general idea of showing respect is by making yourself less, but the manner, whether by bowing the body, kneeling, prostration /uncovering body/ etc etc is matter of custom."—this all applies to *bodily* weakness & inferiority. but now we carry it on to mental inferiority when we do not expect any bodily harm—case of habitual action.— |

33   L'Institut, 1838, p. 340. Mr. Carlyle[157] says that negro certainly has less reasoning powers than Europaean.—Ideots, defective brains. Erasmus does not liken term instinct to muscular movement.—say instinctive *actions, senses, notions,* etc.

Octob. 30th. Dreamt somebody gave me a book in French I read the first page & pronounced each word distinctly. woke instantly but could not gather general sense of this page.—Now when awake I could not picture to myself reading French book quickly, & runing over imaginary words: it appears | as if the mind had dwelt on

34   each word separately, neglecting time, & general sense, anymore than connected with general tendency of the dream.—

It does not hurt the conscience of a Boy to swear, though reason may tell him not, but it does hurt his conscience, if he has been cowardly, or has injured another bad[ly], vindictive.—or lied etc etc |

35   Are the facts (about communication of ideas, etc) of expression lawless, whilst they are the only steady & universal means recog-

nized—no one can say expression was invented to conceal one's thought.—

Macculloch[158] in his Chapter on the Existence of a Deity has an expression the very same as mine about our origin of a notion of a Deity. |

36   We can allow /satellites/ planets, suns, universes, nay whole systems of universes to be governed by laws, but the smallest insect, we wish to be created at once by special act, provided with its instincts, its place in nature, its range, its—etc etc.—must be a special act, or result of laws. yet we placidly believe the Astronomer, when he tells us satellites etc etc The Savage admires not a steam engine, but a piece of colored glass, is lost in astonishment at the artificer.—Our faculties are more fitted to recognize the wonderful structure of a beetle than a Universe.—[159] |

37   November 20th Saw the youngest child of H. W.[160] constantly when refusing food, turn his head first to one side & then to other, & hence rotatory movement negation.—he dropped his head when he meant to eat, hence assertion.—but nodding is less strongly marked than negation

Marianne[161] says, that she has constantly observed that very young children express the greatest surprise at emotion in her countenance,—before they can have learnt by experience, that movements of face are more expressive than movements of fingers. —like kitten with mice— |

38   A person with St. Vitus' dance badly, told should have shilling to walk to door without touching table—cannot avoid it.—curious mixture of voluntary & involuntary movements.—

Person with sore-throat told not [to] swallow spittle, will have involuntary flow & desire to swallow.—tells himself not to turn in bed, will turn in bed.—In case spittle, effect of thought is to make

39   saliva flow, & therefore thinking of subject, even when | wishing not to flow—flow it will.—[162]

My father told Miss C. of the bad conduct of Mrs. C (her brother's wife) & she said nothing but *shrugged* her shoulders.— analyse this.—Miss C. quite aware & indignant with Mrs. C. but had no influence over her.—

Hensleigh says Douglas /& Spencer/, an old Scotch Poet, has numerous lines of poetry.—sounds singularly adapted to subject see <A> I think this argument might be used to show language had a beginning, which my theory requires. |

40   There probably is some connection between very limited reasoning powers & the fixing of habits, for instance the Birgos[163] opening a Cocoa nut shell at one end.—Children & old people get into habits.—we probably can hardly form an idea of a mind so limited as Birgos to become absorbed by one end of Cocoa nut.— |

41   November 27th. Sexual desire makes saliva to flow /yes, *certainly*/ curious association: I have seen Nina licking her chops.—someone has described slovering teethless-jaws as picture of disgusting lewd old man. ones tendency to kiss, & almost bite, that which one sexually loves is probably connected with flow of saliva, & hence with action of *mouth* & jaws.—Lascivious women are described as biting: so do stallions always.—No doubt man has great tendency to exert all senses, when thus stimulated, smell, as Sir C. Bell[164] says, & hearing music, to certain degree sexual.—The association of saliva is probably due to our distant ancestors having been like *dogs* to bitches.[165]—How comes such an association in man.—it is bare fact, on my theory intelligible |

42   An habitual action must some way affect the brain in a manner which can be transmitted.—this is analogous to a blacksmith having children with strong arms.—The other principle of those children which *chance* produced with strong arms, outliving the weaker ones, may be applicable to the formation of instincts, independently of habits. The limits of these two actions either on |

43e   form or brains very hard to define.—Consider the acquirement of instinct by dogs, would show habit.—

    Take the case of Jenner's <Hyaena> Jackall,[166] an animal not destined by nature to exist & carrying /like other hybrids/ with <the> it the provision for death.—can we deny that brain would be intermediate like rest of body? Can we deny relation of mind & brain. ((Do we deny that the mind of a greyhound & spaniel differ from their brain?)) then can we deny that the grandchild dug for mice from some peculiarity of structure of brain?—is this more wonderful than memory affected by diseases, etc, etc, double consciousness? What other explanation—can we suppose some essence. |

44e   The facts about crossing races of dogs on [the basis of] their instincts, *most important,* because they obey the same laws, as the crossing of jackall & Fox & wolf & dog.—the only test this is most important: can there be stronger analogy than the tendency to hybrid greyhound to hunt hares /& leave the sheep/ & jackall to skulk about & hunt mice /Jenner's Jackall/—Have we [illegible] right to deny identity of instincts.—O[167]

    No one doubts that a cross of bull dogs increase the courage & staunchness of greyhounds.—bull-dogs being preferred from not having any smell.—[168] |

45   28th. November.— Think, whether there is any analogy between grief & pain—certain ideas hurting brain, like a wound hurts body —tears flow from both, as when one burns end of nose with a hot razor. Joy a mental pleasure. with pleasure of senses. The shudder of pleasure from pleasure of music.

Audubon[169] IV Vol. of Ornith, Biog. case of Newfoundland dogs who will not enter water, till he sees whether birds badly wounded, or only winged.—fetches two birds out at once.— |

46   *Old* people (Antiquary Vol. II, p. 77) [170] remembering things of youth, when new ideas will not enter, is something analogous to instinct, to the permanence of old hereditary ideas.—being lower faculty than the acquirement of new ideas.—

Walter Scott (Antiquary), Vol. II, p. 126, says seals knit their brows when incensed.—

A Dog may hesitate to jump in to save his master's life,—if he
47   meditated on this, it would be conscience. A man might not | do so even to save a friend, or wife.—yet he would ever repent, & wished he had lost his life in doing so.—nor would he regret /having acquired/ this sense of right (whether wholly instinctive as in the dog, or chiefly habitual as in man), for it added much to the happiness of his life, & the chance of so dreadful a consequence to each man is small.

Man's intellect is not become superior to that of the Greeks (which seems opposed to progressive developement) on account of dark ages.—Look at Spain now.—Man's intellect might well deteriorate.— ((effects of *external* circumstances)) ((In my theory there is no absolute tendency to progression, excepting from favorable circumstances!)) [171] |

48   We must believe, that it requires a far higher & far more complicated organization to *learn* Greek, than to have it handed down as an instinct.—Instinct is a modification of bodily structure / (connected with locomotion) / ((no! for plants have instincts)) /either/ to obtain a certain end; & intellect is a modification of <intellect> /instinct/—an unfolding & generalizing of the means by which an instinct is transmitted.— |

49   Arguing from man to animals is philosophical, viz. (man is not a cause like a deity, as M. Cousin says),[172] because if so ourang-outang,—oyster & zoophyte: it is (I presume, see p. 188 of Herschel's Treatise) [173] a "travelling instance" a "frontier instance," —for it can be shown that the life and will of a conferva[174] is not an antagonist quality to life & mind of man.—& we do not suppose an hydatid to be a cause of itself.—[by my theory no animal as now existing can be cause of itself.]CD & hence there is great probability against free action.—on my view of free will, no one could discover he had not it.— |

50   The memory of Plants, must be association,—a certain round of actions take place every day, & closing of the leaves, comes on from want of stimulus, after certain other actions, & hence become associated with them.—The establishment of this principle of Association will help my theory of sensitive Plants.[175] |

51   Habitual actions, (independent of mind) in the intestinal functions etc etc etc.—bears the same relation to true memory, that the formation of a hinge /in a bivalve shell/ does to reason.[176]—an inflamed membrane from local irritation to passion.—

Blushing is intimately concerned with thinking of ones appearance—does the thought drive blood to surface exposed, face of

52   man, face, neck, /upper/ bosom in women: like erection | shyness is certainly very much connected with thinking of oneself.— /blushing/ is connected with sexual, because each sex thinks more of what another thinks of him, than of any one of his own sex— Hence, animals, not being such thinking people, do not blush.— sensitive people apt to blush.— —The power of vivid mental affection, on separate organs most curiously shown in the sudden cures of tooth ache before being drawn.— |

53   My father /even/ believes that the general talking about any disease tends to give it, as in cancer, showing effect of mind on individual parts of body. (If you <think> /fear/ you shall not have e—n, /or wish extraordinarily to have/ you wont.)— —No surer way to blush, than particularly to wish not to do so.—How directly personal remark will make any one blush.—Is there not some saying about a person even blushing in the dark—/so modest a person/. A person who blushes in the dark is proverbially a most modest person. |

54   One carries on, by association, the question, "one [i.e., what] will anyone, especially a woman think of my face?" to ones moral conduct.—either good or bad, either giving a beggar, & expecting admiration, or an act of cowardice, or cheating.—one does not blush before utter stranger, or habitual friends,—but half & half. Miss F. A. said to Mrs. B. A., how nice it would be if your son would marry Miss O. B. Mrs. B. A. blushed. Analyse this:— |

55   Let a person have committed any /concealed/ action he should not, & let him be thinking over it with sorrow,—let the possibility of his being discovered by anyone, especially if it be a person, whose opinion he regards, <see> feel how the blood gushed into his face,—"as the thought of *his* knowing /it/, suddenly came across her, the blood rushed to her face."—One blushes if one thinks that any one suspects one of having done either good or bad action, it always bears some reference to thoughts of other person. |

56   Decémb. 27th. Fear loose[ns] the sphincter muscles, only on the principle that it paralyzes all muscular action ((like does an injury of the Spine—)) ((in man and animals)) Blubbering of a child (different in different ones?) ((in the most *perfect* fainting, sphincters are loosed)) is a convulsive action to remove disagreeable impression like true convulsion. (Hence passes into convul-

sion?) squeeze out tears replaced & squeezed out again—as power of mind by habit gets more perfect over voluntary muscles, these |

57　convulsive actions—(except in weak people & hysterical people inclined to convulsive actions) .—But the lachrymal gland is /not/ under voluntary power, (or only very little so), & hence by association, there pour out tears, & there is slight convulsive wrinkling of some of the muscles /or *twitching*/.—But why does joy, & *other emotions* make grown up people cry What is emotion?

At end of Burke's[177] essay on the sublime and Beautiful there are some notes, & likewise in Wordsworth's dissertations on Poetry.— |

58　From the manner short-sighted people frown, frowning must have some relation to short-sightedness.— ((The expression of shame-facedness for shyness, having been invented, prove [proof?] of the difference, which my theory believes in.—)) do not short-sighted people squinny when they consider profoundly,—this will be curious if it is so.—frown with grief, ?bodily pain? frown *shows the* mind is *intent on one object.—*

With respect to my theory of smile, remember children smile before they laugh.—Has frowning anything to do with ancient movement of ears— |

59　A man shivers from fear, sublimity, sexual ardour.—a man cries from grief, joy & sublimity.

January 6th.

What passes in a man's mind, when he says he loves a person— do not the features pass before him marked, with the habitual expression of those emotions, which make us love him, or her.—it is blind feeling, something like sexual feelings—love being an emotion does it regard /is it influenced by/ other emotions?

When a man keeps perfect time in walking, to chronometer, is seen to be muscular movement. |

60　The *Blushing* of Camelion & Octopus strong analogy with my view of blushing—in former irritation on a piece of skin cut off made the blush come.—it is an excitement of surface *under the will?* of the animal.—

Jan. 21, 1839. Herschel's Discourse p. 35.[178] On origin of idea of causation (succession of night & day does not give notion of cause) : d[itt]o p. 135.[179] on the importance of a name with reference to origin of language

My father says old people first fail in ideas of time, & perhaps of

61　space, in latter respect he thinks | he certainly has observed that some people of very weak intellect (as Miss Clive) [180] have only possessed very loose ideas. —Have children loose ideas of time?— Characteristic of one kind of intellect is that when an idea once takes hold of the mind, no subsequent ones modify it.— ((Weak people say I *know* it, because I was always told so in childhood, hence the belief in the many strange religions.))

Emma W.[181] says that when in playing by memory she does not think at all, whether she can or cannot play the piece, she plays better than when she tries is not this precisely the same, as the double-conscious[ness] kept playing so well— |

62   L[rd]. Broughham[182] /Dissert./ on subject of science connected with Nat. Theology.—says animals have abstraction because they understand signs.—very profound.—concludes that difference of intellect between animals & men only in kind. probably very important work.

Feb. 12, 1839 Sir H. Davy Consolats: "the recollections of the infant likewise before two years are soon lost; yet many of the habits acquired in that age are retained through life." p. 200.[183]—"The desire of glory, immortal fame, etc so common in the young are symptoms of the infinite & progressive nature of intellect, indication of better life. p. 207.[184] |

63   March 16th. Is not that kind of memory which makes you do a thing properly, even when you cannot remember it, as my father trying to remember the man's Christian name.[185] writing for the surname, analogous to instinctive memory, & consequently instinctive action.—Sir J. Sebright[186] has given the phrase "hereditary habits" very clearly, all I must do is to generalize it, & see whether applicable at all cases.—& analogize it with ordinary *habits* that is my new part of the view.—let the proof of hereditariness in habits be considered as grand step if it can be generalized.— |

64   The tastes of man, same as in Allied Kingdoms—*food, smell* (ourang-outang) , *music,* colours we must suppose Pea-hen admires peacock's tail, as much as we do.—touch apparently, ourang outang very fond of soft, silk-handkerchief—cats & dogs fond of slight tickling sensation.—in savages other tastes few.

March 16th. Gardiner's Music of Nature.[187] p. 31, remarks children have no difficulty in expressing their wants, pleasures or pains

65   long before they can speak, | or understand—thinks so it must have been in the dawn of civilization—thinks many words, roar, scrape, crack, etc, imitative of the things.—CD[I may put the argument that many learned men seem to consider there is good evidence in the structure of language, that it was progressively formed (names like sounds) . Horne Tooke's[188] tenses, etc etc—if so & seeing how simple an explanation it offers of radical diversity of tongues.— |

66   [Emotions are the hereditary effects on the mind, accompanying certain bodily actions.]CD ?but what first caused this bodily action, if the emotion was not first felt?— ((without /abrupt/ flush, acceleration of pulse, or rigidity of muscles,—man cannot be said to

be *angry.*)) ((He may have pain or pleasure these are sensations))

<Gardner in his work> In the life of Hayd & Mozart, fine music is evidently considered analogous to glowing conversation of several people.[189]

Children have an uncommon pleasure in hiding themselves & skulking about in shrubbery, when other people are about: this is analogous to young pigs hiding themselves; a hereditary remains
67 of savages state.— | N.B. According to my view marrying late, will make average of life longer.—for short-lived constitutions will then be cut off.

<Horses> Colts cantering in S. America capital instance of hereditary habit: there must, however, be a *mental impulse* (though unconscious of it) to move its legs so, as much as in the young salmon to go towards the sea, or down the stream; which it does unconsciously of any end.—NB. There is wide difference, between the means by which an animal performs an instinct, & its impulse to do it.—[the means must be present on any hypothesis whatever]CD an animal may so far be said to *will* to perform an instinct that it is uncomfortable if it does not do it.— |

68 My theory explains how it comes that the heart is the seat of the emotions,—but are not love & hate emotions; what are their characteristics;—they are more truly sensations?? a kind of mental pain & pleasure.—

The Revᵈ Algernon Wells[190] Lecture on animal instinct, 1834: p. 15, "To act from instinct is to be guided to the performance of a number of pre-arranged actions, which will bring about a certain result, while the creature performing those actions, neither knows nor intends the result they will effect." ·this not wholly true, for we must grant a bird knows what is about when building its
69 nest; it knows its object but not *result* | (?first time of building?), but not the means of performing it.— p. 14. There is scarcely a faculty in man not met with in the lower animals.—hence the general aim of fables, & expressions are cunningness of fox, industry of bee etc etc.[191]— p. 15. "instincts act with *unerring* precision." —no[192] p. 17. Contrasts the *invariability* of *instinctive* powers in individuals of the same species with *variability* of *reasoning* power in one species man.—false instinctive pointing varies.[193]— p. 18. Animals possess strong imitative faculty: pure instinct is not imitative: imitation seems invariably associated with reason:[194] [NB. insects which have never seen their parents offer best cases of instincts].CD All this may be true, but relation of imitation & reason must be thought of.[195]— |

70 p. 19. animals capable of education (this is again assumed as more allied to reason than instinct.)[196] Mr. Wells I can see mentally refers by reason *knowledge gained by reason:* & then these

qualities of imitation & education may be used as argument.—for *instinctive knowledge* is not gained by instruction, or imitation.[197]— p. 20. Animals may be called "creatures of instinct" with some slight dash of reason so men are called "creatures of reason," more appropriately they would be *"creatures of habit."* CD[as the bee makes its cells, by means of ordinary senses & muscles, we cannot look at him, as machine to make cell of certain form. (& especially as it adapt[s] its cell to circumstances), it must have

71 impulse to make | a cell in certain way. which way its organs are sufficient for hence it must some way be able to measure the cell;[198]

    p. 22. Instincts & structures always go together:[199] thus woodpecker:[200] but this is not so, the instincts may vary before the structure does; & hence we get over an apparent anomaly, for if anyone has taken the woodpecker as an example fitted for climbing, his arguments partly fall, when a species is found which does not climb. Instinct may be divided into migration,—subsidiary to *food* & temperature /molting & breeding/ instincts, sexual, social /subordinate to/ self preservation, (knowledge of enemies). use of muscles, progression,—use of senses,—knowledge of location ducks & turtles running to water.—young crocodiles snapping— |

72   p. 28. how curious the means of guiding themselves through the air,—waterbirds, the bee to its nest,[201]—cats when carried in confinement,—carrier pidgeons proverbially carried to long distance in dark "it is inspiration."[202]—this is class of so called instincts to which my theory no way applies.—it is the acquirement of a new sense,—bats avoiding strings /in the dark/ as well might be called instinct,—migrating to one spot, this is indeed instinct—Australian man, may be called instinctive: the facts of the memory of roads long after once visited by horse & dogs. (even blind horses & dogs)

73 shows it is somewhat analogous | to memory.—

    Shrugging shoulders seems sign of helplessness

    E.[203] says she can perceive sigh, commences as soon as painful thought crosses mind, before it can have affected respiration

    V. E.[204] p. 125. Wrong Entry.[205] |

73–74   [lower half excised, not found]

74   Madagascar Lemur seemed to LIKE *Lavender* WATER /very much/ Henslow.[206]

    M[rs] Necker[207] has remarks on the means by which children learn (probably not only experience, but also /by an/ instinct which is only present in youth (Mem: Mr. Worsley's story of chicken) to know that which we touch & what we [excised:] the same. (this Hensleigh [excised]

75 therefore problem is how we know that thing is same, which

touches two parts of our bodies /or touches one part very quickly successively/,—[& we know from experiment of crossing fingers that we only do know that it is one. when applied in peculiar manner.—]CD

75e April 3ᵈ, 183[9]

The Giraffe kicks with front legs & knocks with back of Head, yet never puts down its ear, good to contrast with horses, asses, Zebras etc etc.—Here there is kicker but not bite.208— |

75–76 [lower half excised, found]

76 Henslow remarks that Chimpanzee pouted & whined, when men went out of room.—

all theories of magnetic power in birds, seeing the sun, etc are
76e absolutely useless when applied to bird, which | have been carried in hampers if they have not known the direction in which they STARTED, they cannot return.—Hence I conclude pidgeon taken
77 little way, *whirled* & then taken other way, would not find209 | its way back??—this is not instinct, but a faculty or sense.—We know not how stonge henge raised, yet not instinct, but if all men placed stone in same position, it would be instinct—instinct is hereditary knowledge of things which might be /possibly/ acquired by habit. so bees in building cells, must have some means of measuring cells, which is a faculty, they use this faculty instinctively; watchmaker has faculty by his instruments to make toothed wheel.
78 he might by his instinct make watch, but he does it by | reason & experience or habit.—so bird migrating to certain quarter is instinct, but his knowledge of that quarter is faculty, whether by sun & heavens, or magnetic virtue.—the most probable supposition with respect to pidgeons, is that they do know from look of Heavens points of compass, & they do know which way they go; & so return.—/but does not apply to dogs.—/ they may do all this instinctively /yes, because power varies in breeds/, something of kind oneself knows in walking [one feels inclined to stop at right
79 number of house though one cannot remember it.]CD | back, without consciousness & by *habit*. such habit of knowledge of points of compass may be instinctive.

it is a test to know how much of the wonder consists in the action being performed or emotion felt in early childhood (before experience is habit) could be formed or afterwards.—child sucking whole wonder instinctive—carrier pidgeon just as wonderful in old bird as new.—migration, only more wonderful in young, because cannot have been taught, where to go—the act of crossing
80 the | sea in dark night & not loosing its direction equally wonderful in young & old.—These facts point out some essential difference, which clearly ought to be separated—We apply instinct to

one part or another—but (an *instinctus* means *stained* in?) had better refer to to the hereditary part of it.—& *faculty* (faculty being always hereditary helps this confusion.—) |

81   Hensleigh considers breathing instinctive, certain heart beating may be considered also such.—hereditary habit is a part never subject to volition,—like plants going to sleep.—

"A bird has the faculty of finding its way, which in certain species is instinctivedly /not least by experience/ directed to certain quarter."—"An animal has faculty of walking, which in man is learnt by experience is in others is acquired instinctively." So

82   with sight,—so a Bee has the faculty of | building /regular/ cells —[but this faculty probably is instinctive, namely the knowledge of size is merely judged by eye, & use of limbs, etc, or it result from mere impulse to save wax]^{CD} which it instinctively exerts in concert with others in building comb—My faculty often will turn out to be instincts, & so in some senses, is sight. [The faculties bear so close a relation to the senses, that one feels no more surprise at it &.feels no more inclined to ask |

83–86e [excised, not found]

87e   if dislike, distaste, & disapproval were not something more than the unfitness of the objects then viewed, organs adapted to other objects, (as that senna is necessarily disagreeable to organs adapted to like sugar, acid, etc, which may be doubted for possibly even taste of senna, might be acquired, as the Turks[210] have of Rhubarb: again on other hand, it is said people, who like sweet things dislike others.—dogs dislike perfume) I should think, great principle of liking, was simply *hereditary habit*.— |

88e   A blind man might be born with the idea of scarlet, as well as remember it.—

Why do children pout & not men—orang-outang & chimpanzee, pout.—Former whines just like a child.

Get a dictionary & make list of every word expressing a mental <desire> quality etc etc. |

89   Mackintosh[211] Ethics

p. 97 on Devotional feeling

p. 103—Abstraction

p. 152 Perception: very different from *emotion*.—The former is used with regard to the senses. Reason does not lead to action.

p. 248. Theory of Association, owing to time when entered brain, by contiguity of parts of Brain. Mackintosh first clearly insisted on assoc of ideas & emotions. ?rather ideas & bodily actions make the emotions.—

p. 272. Some remarks applicable to my theory of happiness.

Bell[212] on the Hand.

p. 191. Says <childr> babies have an instinctive fear of falling, &

p. 193 that they perceive the difference on being carried up or downstairs, or dangled up & down—in latter case they struggle
90 their arms.— | do. p. 306. "The eyes are rolled upwards during mental agony, & whilst strong emotions of reverence & piety are felt." It appears to me mere consequence of stooping, as sign of humility.—

I suspect very strong argument might be advanced, that animals have reason, because they have memory.—what use this faculty if not reason.—or does this reasoning apply chiefly to recollection, yet a dog hunting for a bone shows he has recollection.—

Lamarck.[213] Phil. Zoolog. Vol. II, p. 445. If we compare the judgments & actions of a young animal with an old (dog, horse, sow), we perceive great difference (& is not this difference same, but less in degree, as between man & child—) —what differs—Not<reason> instinct, for its character is invariability,—if explained by habits, useful to itself, how gained. reason? Or some unnamed faculty— |

91 Lamarck.[214] Philosop. Zoolog. /p. 284, Vol. II./—gives explanation & instance of *starting* identical with mine.—Lamarck,[215] Vol. II. p. 319. Habits more prevalent in proportion to intelligence less.—

p. 325 to 29.[216]—Habits becoming hereditary from the instincts of animals.—almost identical with my theory, no facts, & mingled with much hypothesis.—See M.S. notes where strong argument in favour of brain forming the instincts,—could brain make a tune on the piano-forte, yet if every individual played a little (& something destroyed bad brain.) [217] |

92 See p. 90.—The relation of reason to organs of locomotion—or that our faculties have been given us to exist, is clearly seen in the absurdity of a tree having reason: or dog having high powers, without hand & voice.—there is some great puzzle in what Sir J. M.[218] says of pure reason not leading to action, & yet our emotions being only bodily actions associated with ideas.—

A sigh, is an abortive groan, more power over muscles of voice than respiration,—like sigh before false sneeze:— |

93 "A Dissertation of the Influence of the Passions"[219]—
p. 37. The increase of Biliary secretion attends passion.
p. 39. The sweat that accompanies fear is the same, as that which attends great weakness—<diarrhaea> & syncope
p. 42. Sighing from grief is method of increasing languid circulation—no, for sighing comes on before circulation is affected.
p. 44. Jealousy causes spasm in bile duct, & throws bile in circulation.
p. 75. Haller says toothache even from carious tooth cured by sight of instrument.— |

94   Bennett's[220] Wanderings, Australian Dog does not Bark—quotes
Gardner's[221] Music of Nature to show barking not natural. (Vol. I.
p. 234)

Vol. II, p. 153 /d[itt]o/,[222] an account of a monkey in a passion
like Jenny.—Dr. Abel[223] has given an account of an Ourang.—see
his Travels.—

When one sees in Cowper[224] whole sentences spoken & believed
to be audible, one has good ground to call imagination a faculty,
a power, quite distinct from self—or will |

95–96   [excised, not found]

97   Mr. Hamilton[225] on vital laws (in the Athenaeum Library)
describes effects of emotions—fear giving goose skin—& hair standing on end.—

July 20th Intelligent Keeper, Zoology Gardens told me he has
often watched tame young wolf & it never dropped its ears like
dog—wagged its tail /a little/ when attending to anything or excited.—so do young dingos, as I saw, wag tail when watching any-

98   thing—Keeper does not think | they drop their ears.— —

George the lion is extraordinarily cowardly.—the other one
nothing will frighten—hence variation in character in different
animals of same species.— |

99   The general (as I believe) contempt at suicide (even when no
relatives left to lament) is owing to the feeling that the instinct of
self-preservation is disobeyed—I often have (as a boy) wondered
why *all abnormal* sexual actions or even impulses (where sensations of individual are same as in normal cases) are held in abhorrence. It is because instincts to women is not followed; good
case of instinctive[226] |

99–100   [upper half excised, not found]

100   . . . not eating cause disgust, because he does not go *against instinctive feeling* only does not fullfil, like continent man.—a man
eating what others by *habit* (not instinct) think not fit, as can-
nabalism, is held in *abhorrence.*—all this makes analogy of actions
with & against benevolent & parental instincts very clear, even to
the cold or benevole[nt] continent man |

101   Hume[227] has section (IX) on the Reason of Animals, Essays,
Vol. 2 ((Sect. XV. Dialogue on Natural Religion.[228]— also on
origin of religion or polytheism, at p. 424, Vol. II however he
seems to allow it is an instinct.))

I suspect the endless round of doubts & scepticisms might be
solved by considering the origin of reason, as gradually developed.
See Hume[229] on Sceptical Philosophy.

Hume[230] has written "Natural Hist. of Religion" on its origin in
Human Mind.

Andrew Smith[231] says hen doves & the *female chameleon* court the males by odd gestures. |

102   In one of the six (?) first Vol. of Silliman's Journal[232] paper showing that the signs invented for Deaf & Dumb school & used between Indian tribes are Many the same.— |

103   Philosoph. Transactions Vol. 44, 1746–47. Paper like Sir Ch. Bell on Expression /First Crounian Lectures by Parsons/,[233] following pages contain remarks worthy of attention, p. 15, 25, 40, 61 [CD][a person is here said to open mouth in fright because nature intends to lay open all senses: Horse pricks his ears (& snort clears nostrils) when frightened, does not hare & rabbit depress them from squatting.—p. 64. closing both eyelids express[es] contempt; p. 76. children have been tickled into excessive laughter & so into convulsions.—Paper must be referred to, if I follow up this subject & a reference to Brun's[234] work.—

Shutting eyes in *contempt* opposite action to opening eyes in *fear* |

104   The effect of habitual movements in muscles of face. is well seen in short-sighted people.—hence origin of expression— |

105   There are some instincts unintelligible, in the end gained /& therefore the/ cause, and origin being so is not odd; for instance wild cattle & deer pursuing a wounded one,—porpoises a ditto—it is probably some secondary one—blood being disagreeable & anything disagreeable being pursued.—

A dog turning round & round is some old instinct handed down & down. Mem. Nina used to get into hay & make a nest for herself—the object is to make saucer-shaped depression.— |

106, 108   [blank]

107   Does music bear any relation to the period when men communicated before language was invented.—were musical notes the language of passion & hence does music now excite our feelings.[235]

How does Social animal recognize /& take pleasure in/ other animals, (especially as in some <instincts> insects which become in imago state social) by smell or looks. but it does not know its own smell or looks, & therefore there must be some instinctive feeling which is pleased by other animals smell & looks.—no doubt it may be attempted to be said that young animal learns parent

109   smell & look & so by association receives pleasure. This | will not do for insects.—if this view holds good, then man, a socialist, does not know other men by smell, but by looks hence some obscure picture of other men, & hence idea of beauty.—The social affections of animal taking man in place of other animals is hostile /is subversion of/ to this view, & fowls hatching stones, in

some degree is so.—ideas of beauty of music are great distinguishing character between man & animals.— |

110 [blank]

111   Double consciousness only extreme step of an ideal argument held in one's own mind, & Dr. Hollands story of man in Delirium tremens he'aring other man speak, shows, that consciousness of personal identity is by no means a necessary part of man's mind.—

At Maer Pool I saw many coots & waterhens feeding on grassy bank some way from water, *suddenly,* as if by word of command, they all took flight & flappered across pool to bed of flags I was astonished & having looked round saw at considerable distance a

112   very large | hawk which are /so/ rare here, that probably few had ever before seen one, yet all flew to bed of flags. Hernes are common, not unlike in size in the air at a distance. How can such an instinct arise?? ((It would appear that an instinct long remains if no steps are taken to eradicate it.)) ((Emma says her tame rabbits were not frightened at a dog.—))

The instinct against man is perhaps as strong as against hawk, but the birds at Maer have learned that he is not dangerous. wildducks would have fled equally if man had appeared, though instinct so firmly implanted, birds soon learn to disobey it—I have seen hawk & sparrow in Shrewsbury garden picking from same bone. |

113   A child born on the 1st March was frightened on the 24th of May at Cresselly[236] by the boys making faces at it, so much so that the nurse had to carry it out of the room; nearly 3 months old.

What is absurdity, why does one laugh at it—

Sensation of disgust with nausea (when stomach a little disordered) at thought of almost anything ugly baby association—pouting child same as anger, lips not compressed, sullen, protruded, determined to do nothing, & so manifesting sullenness. |

114 [blank]

115e   Circumstances having given to the Bee its instinct is not <more> less wonderful than man his intellect.

Lyell has seen a little dog go to the assistance & bite a big dog. which was fast struggling with another large dog his companion.

Monkeys /Olgleby/[237] remember with distress their companions —a /blue [?]/ Gibbon whose companion had been dead about two months. saw a /black/ spider monkey brought at it opposite end of house & commenced a most lamentable howl & was not comforted until the Keeper took it in his arms & carried to see.—[238] |

116e [blank]

117   A Dog /whilst/ dreaming, growling & yelping, /& twitching paws/ which they only do when considerably excited, shows their power of imagination—for it will not be allowed they can dream & not have day-dreams.—think well over this; it shows similarity

in mind.—think of Eyton's[239] horses becoming white with <foam> & sweat, when hearing merely hunting horn—association or imagination |

118 [blank]

119–120e [excised, not found]

121 Ernest W.[240]—playing with Snow[241] when 2-1/2 years old, was frightened when Snow put a gauze over her head, & came near him, although knowing it was Snow.—Is this part of same feeling which makes us think anything ugly—a beau-ideal feeling. Same effect as acting on us. ((Effie Wedgwood[242] <The Baby>, April 28, 1840 was frightened at wild beasts in Zoology Garden)) |

122 [blank]

123–124e [excised, not found]

125 A child crying, frowning, pouting, smiling just as much instinctive as a bull calf, just born butting, or young crocodile snapping. —these I think are better instances of instinct (highly useful as only means of communication) in man, than sucking—[I assume a child pouts who has never seen other pout][CD] |

126 [blank]

127 Goldsmith's[243] Essays, No. XV on sounds of words being expressive (Vol. 4 of Works) |

183 Ourang do not move eyebrows.—or skin of head, Cyanocephalus macacus. *Cercopithecus?* very much. (Keeper says some of the monkeys move the ears but Chimpanzee does not gradation towards man.) Macacus especially pulls back skin of whole forehead & ears.—emotions of every kind. [Are monkeys right-handed??][CD] Cyanocephalus, Macacus, Niger, Cercopithecus make labial st. st. S. American monkeys pull back skin forehead very little. |

Does blood go in <body> face in passion? cry? |

184 "Adam Smith[244] Moral Sentiments" much on life & character "Humes[245] Dissertations on the Passions." "Hartley"[246] I should think well worth studying—

"Thomas Brown" on Association.[247] Worthy of close study.— full of practical observations. |

185? Do people of weak intellects easily fall into *habits*. Get facts about instincts of mongrel dogs

Do blubbering children, if of convulsive tendency easily fall into convulsions.

A carrier pidgeon carried & turned round & round in fainting state would it then know its directions.—

In slight convulsions, are the muscles of the face first affected?

Can shivering & trembling be considered convulsive.—is convulsion an involuntary movement of voluntary muscles—if so what is trembling palsy? |

128. Waterton, Charles, "The Habits of the Carrion Crow," *Magazine of Natural History, and Journal of Zoology, Botany, Mineralogy, Geology, and Meteorology,* Loudon, London: 6:208–214, 1833, p. 212.

129. Berth Hill, 8½ miles northwest of Shrewsbury.

130. Waterton, Charles, *Essays on Natural History, Chiefly Ornithology,* etc., London, 1838, p. 292.

131. Pincher and Nina, pet dogs.

132. Double vertical lines at the beginning and end of this statement in the notebook.

133. Probably Worsley, Rev. T., M.A., Master of Downing College, Cambridge University, Vice Chancellor of the University, F.G.S., and member of the Athenaeum.

134. Lavater, *op. cit.,* Vol. 3, Part II, pp. 297–298.

135. *Ibid.,* Vol. 3, Part I, p. 21: "The nose is the seat of derision, its wrinkles contemn. The upper lip when projecting speaks arrogance, threats, and want of shame: the pouting under lip ostentation and folly."

136. *Ibid.,* Vol. 3, Part II, p. 397: "Fleshy lips have always a struggle to maintain with sensuality and indolence."

137. *Ibid.,* Vol. 3, pp. 37–38: "Miscellaneous Quotations. 1. From Burke, on the Sublime and Beautiful. 'Campanella had not only made very accurate observations on human faces, but was very expert in mimicking such as were any way remarkable . . . he was able [thereby] to enter into the dispositions and thoughts of people as effectively as if he had been changed into the very men. . . . Our minds and bodies are so closely and intimately connected, that one is incapable of pain or pleasure without the other.' "

138. Malthus, T. R., *An Essay on the Principle of Population; or, a View of its Past and Present Effects on Human Happiness; with an Inquiry into our Prospects Respecting the Future Removal or Mitigation of the Evils which it Occasions,* 6th ed., 2 vols., Murray, London, 1826 Vol. 1, p. 31: "Women treated in this brutal manner must necessarily be subject to frequent miscarriages, and it is probable that the abuse of very young girls, mentioned above as common [in New Holland], and the too early union of the sexes in general, would tend to prevent the females from being prolific." P. 32: "Women obliged by their habits of living to a constant change of place, and compelled to an unremitting drudgery for their husbands, appear to be absolutely incapable of bringing up two or three children nearly of the same age."

139. *Ibid.,* Vol. 2, p. 256: "Natural and moral evil seem to be the instruments employed by the Deity in admonishing us to avoid any mode of conduct which is not suited to our being, and will consequently injure our happiness. If we are intemperate in eating and drinking, our health is disordered; if we indulge the transports of anger, we seldom fail to commit acts of which we afterwards repent; if we multiply too fast, we die miserably of poverty and contagious diseases." Pp. 263–264: "It may be further remarked . . . that the passion is stronger, and its general effects in producing gentleness, kindness, and suavity of manners, much more powerful, where obstacles are thrown in the way of very early and universal gratification . . . in European countries, where, though the women are not secluded, yet manners have imposed considerable restraints on this gratification, the passion not only rises in force, but in the universality and beneficial tendency of its effects; and has often the greatest influence in the formation and improvement of the character, where it is the least gratified." P. 264: ". . . much evil flows from the irregular gratification of it [i.e., the passion between sexes]. . . ."

140. Comte, *op. cit.,* p. 280.

141. Jenny, an ourang-outang at the Zoological Society Zoo, London. Barlow, Nora, *Charles Darwin and the Voyage of the Beagle,* Pilot, London, 1945, pp. 147–148.

142. Baboon (*Cynocephalus*, Cuvier).

143. Whewell, William, *History of the Inductive Sciences, from the Earliest to the Present Times*, 3 vols., Parker, London, 1837, Vol. 1, p. 334.

144. (Added in pencil.) Fuegia Basket and Jemmy Button were Fuegians returned to Tierra del Fuego by Capt. FitzRoy and the *Beagle* during Darwin's voyage.

145. Scrope, G. P., *Memoir on the Geology of Central France; Including the Volcanic Formations of Auvergne, the Velay, and Vevarais, with a Volume of Maps and Plates*, Murray, London, 1827; *Quarterly Review*, 36:437–483, 1827. Passages in this review of Scrope by Lyell to which Darwin must have had reference are: p. 440, footnote, "In short, Mr. Scrope's *elastic vehicle* is the counterpart of Lamarck's *nervous fluid*, that 'subtle and invisible agent,' to which he attributes not only muscular motion, but ideas, sentiment and intelligence. (*Philosophie Zoologique*, Part 3, Chap. 2.) If in attempting to trace back the phenomena of heat, as well as those of the vital functions, we ultimately reach a point which eludes the gross apprehension of our senses, why not unreservedly avow our utter inability to solve such problems?" P. 475, "But the discoveries of astronomy were most pre-eminently beneficial, not so much from their practical utility, although in this respect their services were of inestimable value, but because they gave the most violent shock to the prejudices and long-received opinions of men of all conditions. . . . [Galileo's] sufferings in the cause of truth did not extinguish the spirit of the Inquisition, but gave a death blow to its power, and set posterity free, at least from all open and avowed opposition, to enlarge the Boundaries of the experimental sciences."

146. Scott, Walter: this reference not traced.

147. The words "Descent of Man" are crayoned in above this paragraph.

148. Lutké, Frédéric, *Voyage autour du Monde, exécuté . . . sur la corvette le Séniavine, dans les années 1826, 1827, 1828, et 1829, etc.*, traduit du russe par F. Boyé, Didot, Paris, 4 vols., 2 atlases, 1835–36.

149. The word "Expression" is crayoned in at the top of the page. Also, the words "leave this out not in Library no good" are written across the page.

150. Squib—pet dog at Maer, the Wedgwood home.

151. Probably Stokes, John Lort (1812–1885), Mate and Assistant Surveyor on H.M.S. *Beagle* in 1831. See also references to Stokes's collection of sphaerulites and obsidians, Darwin, C., *Geological Observations*, 2nd ed., Smith, Elder, London, 1876, pp. 71, 79.

152. Reynolds, Joshua, *The Literary Works of Sir Joshua Reynolds to Which Is Prefixed a Memoir of the Author* by H. W. Beechy, 2 vols., Cadell, London, 1835, Vol. 2, pp. 131–132: "[A study of Italian Masters] will show how much their principles are founded on reason, and, at the same time, discover the origin of our ideas of beauty. . . . To distinguish beauty, then, implies the having seen many individuals of that species . . . a Naturalist, before he chose one as a sample [blade of grass] . . . selects as a Painter does, the most beautiful, that is, the most general form of nature."

153. In his copy of *Zoonomia*, Vol. 1 (now in the Cambridge University Library), p. 253, Darwin wrote in pencil in the margin: "tastes hereditary do [ditto]." In the adjacent text Erasmus discusses the role of repetition and imitation in developing concepts of pleasure and beauty: "So universally does repetition contribute to our pleasure in the fine arts, that beauty itself has been defined by some writers to consist in a due combination of uniformity and variety. . . . The origin of this propensity to imitation has not, that I recollect, been deduced from any known principle. . . ." P. 254: ". . . our perceptions themselves are copies, that is, imitations of some properties of external matter; and the propensity to imitation . . . thus constitutes all the operations of our minds." See also Macculloch, John, *Proofs and Illustrations of the Attributes of God*, etc., 3 vols., Duncan, London, 1837, Vol. 3, Chapters: "On the Pleasures provided through the Senses of Odour and Taste; Sense of Seeing, Beauty; Sense of Hearing, Music, and on Pain."

154. From "Everything" to "Bodily" set off by double vertical lines in both the right and left margins.

155. Mayo. Darwin probably has reference to Mayo's comparison of York Cathedral with Fingal's Cave. See OUN6.

156. Edition used by Darwin not traced, but see Reynolds, Joshua, *The Works of Sir Joshua Reynolds, Knt. Late President of the Royal Academy: Containing his Discourses, Idlers, etc.*, by Edmond Malone, 2 vols., Cadell, London, 1797, pp. 150–151: "The general principles of urbanity, politeness, or civility, have been ever the same in all nations; but the mode in which they are dressed, is continually varying. The general idea of shewing respect is by making your self less; but the manner, whether by bowing the body, kneeling, prostration, pulling off the upper part of our dress, or taking away the lower,* is a matter of custom." Footnote: "*Put off thy shoes from off thy feet; for the place whereon thou standest is holy ground. Exodus, iii, 5."

157. Carlyle, H., "Tératologie: Difformités du cerveau," *L'Institut, Journal Général des Sociétés et Travaux Scientifiques de la France et de L'Étranger*, 5:340, 1838.

158. Macculloch, *op. cit.*, Chapter IV, pp. 94–127, "On the Existence of the Deity. Nature of Proof. Sources of Belief." P. 95: "To proceed a further step, somewhat more rapidly than metaphysics do, the proof of the existence of a Supreme Creator depends therefore on our belief in a cause, or in what has been termed causation."

159. Darwin's reference to the savage and the steam engine may have had its origin in the following: Davy, Humphry, *Consolations in Travel, or the Last Days of a Philosopher. With a Sketch of the Author's Life, and Notes*, London: John Murray, 1830. "A savage who saw the operation of a number of power-looms weaving stockings, cease at once on the stopping of the motion of a wheel, might well imagine that the motive force was in the wheel; he could not divine that it more immediately depended upon the steam, and ultimately upon a fire below a concealed boiler."

160. Hensleigh Wedgwood; the child would have been Ernest Hensleigh, Darwin's nephew, born 1838.

161. Marianne (1798–1858), Darwin's oldest sister; married Henry Parker in 1825, and had had 4 children by this time.

162. See *Zoonomia*, p. 419, for similar statements: "If any one is told not to swallow his saliva for a minute, he soon swallows it contrary to his will, in the common sense of that word. . . . In the same manner if a modest man wishes not to make water, when he is confined with ladies in a coach or an assembly-room; that very act of volition induces the circumstance, which he wishes to avoid, as above explained; insomuch that I once saw a partial insanity, which might be called voluntary diabetes, which was occasioned by fear (and consequent aversion) of not being able to make water at all."

163. Darwin, *Voyage of the Beagle, op. cit.*, 1839, p. 551: "I think this is as curious a case of instinct as I ever heard of, and likewise of adaptation in structure between two objects apparently so remote from each other in the scheme of nature as a crab and a cocoa-nut tree."

164. Bell, Charles, *The Anatomy and Philosophy of Expression as Connected with the Fine Arts*, London: John Murray, 1844: "Nothing sensual is indicated by the form of the human nose; although by depressing it and joining it to the lip, the condition of the brute,—as in the satyr, the idea of something sensual is conveyed."

165. Later Darwin suggested that in dogs licking might have "become associated with the emotion of love" through "females carefully licking their puppies —the dearest object of their love—for the sake of cleansing them." *Expression*, p. 120.

166. Jenner, quoted in Hunter, John, *The Works of John Hunter, F.R.S., with Notes*, James F. Palmer, ed., 4 vols., Longman, London, 1835–1837, 1837, Vol. 4, pp. 329–330: "The following account from Mr. Jenner, of Berkeley, to whom I gave a second remove, viz., three parts dog, is very descriptive of this propensity [i.e., to fall back into original instinctive principles]: 'The little jackal-bitch you gave me is grown a fine handsome animal; but she certainly does not possess the understanding of common dogs. She is easily lost when I take her out, and is quite inattentive to a whistle. She is more shy than a dog, and starts frequently when a quick mo-

tion is made before her . . . her favourite amusement is hunting the field-mouse, which she catches in a particular manner."

167. A circle, i.e., a large O, is drawn here. At the beginning of this page are four words penciled in—"Habits useful to B——" (last word illegible) .

168. See "Questions for Mr. Wynne."

169. Audubon, John James, *Ornithological Biography, or an Account of the Habits of the Birds of the United States of America, and Interspersed with Delineations of American Scenery and Manners,* 5 vols., Adam and Black, Edinburgh, 1831–1849, Vol. 4, 1838, pp. 8–9.

170. This reference to Scott, and the next, have not been traced. See however n. 41.

171. This statement reflects Darwin's early rejection of Progressionism as presented by Erasmus Darwin, *Zoonomia*, p. 505: ". . . all warm-blooded animals have arisen from one living filament, which THE GREAT FIRST CAUSE endued with animality, with the power of acquiring new parts, attended with new propensities, directed by irritations, sensations, volitions, and associations; and thus possessing the faculty of continuing to improve by its own inherent activity, and of delivering down those improvements by generation to its posterity, world without end!" And on page 509, "This idea [viz., THE CAUSE OF CAUSES] is analogous to improving excellence observable in every part of the creation; such as the . . . progressive increase of the wisdom and happiness of its inhabitants. . . ." See also Darwin and Seward, *More Letters, op. cit.,* Vol. 1, p. 41, where Darwin in a letter written January 11, 1844, to J. D. Hooker, said, "Heaven forfend me from Lamarck nonsense of a 'tendency to progression,' 'adaptations from the slow willing of animals,' etc."

172. See Cousin, Victor, *Elements of Psychology; Included in a Critical Examination of Locke's Essay on the Human Understanding, and in Additional Pieces,* transl. from the French, with an Introduction and Notes, by Caleb S. Henry, Hartford: Cooke & Co.,1834, Chapter IV, "Of the Idea of Cause".

173. Herschel, J. F. W., *A Preliminary Discourse on the Study of Natural Philosophy,* Longman, London, 1831, p. 188: "Bacon's 'travelling instances' are those in which the *nature* or quality under investigation 'travels,' or varies in degree; and thus . . . afford an indication of a cause by a gradation of intensity in the effect. . . . The travelling instances, as well as what Bacon terms 'frontier instances,' are cases in which we are enabled to trace that general law which seems to pervade all nature—the law, as it is termed, of continuity. . . . 'Natura non agit per saltum.' "

174. See *Zoonomia,* p. 103 and p. 507.

175. Erasmus Darwin (*Zoonomia,* p. 104) also discusses sensitive plants: "The divisions of the leaves of the sensitive plant have been accustomed to contract at the same time from the absence of light. . . ." In the margin of page 105, in Charles Darwin's handwriting, is, "does habit imply having ideas?" In the adjacent text passage is the following: "And it has been already shewn, that these actions cannot be performed simply from irritation, because cold and darkness are negative quantities, and on that account sensation or volition are implied, and in consequence a sensorium or union of their nerves."

176. Lamarck, J. B., *Zoological Philosophy; An Exposition With Regard to the Natural History of Animals,* Transl., with an Introduction, by Hugh Elliot, Macmillan, London, 1914. [Original: *Philosophie Zoologique,* 2 vols., Dentu, Paris, 1809.] Also discussed the reasoning powers of a bivalve: "There is nothing more wonderful in the alleged skill of the ant-lion (*Myrmeleon formica leo*) which digs out a hole in loose sand and then waits until some victim falls into the bottom of a hole by the slipping of the sand, than there is in the operation of the oyster, which for the satisfaction of all its needs has only to open slightly and close its shell. So long as their organisation remains unchanged they will both continue to do just what they do now, without any intervention of will or reasoning."

177. Burke, *op. cit.* "A Philosophical Inquiry into the Origin of Our Ideas of the Sublime and Beautiful; With an Introductory Discourse Concerning Taste, and Several Other Additions"; Part III, Section XXVII, "The Sublime and Beautiful Compared," and Part IV, Section I, "Of the Efficient Cause of the Sublime and Beautiful."

178. Herschel, *op. cit.*, p. 35: "The first thing impressed on us from our earliest infancy is, that events do not succeed one another at random, but with a certain degree of order, regularity, and connection;—some constantly, and, as we are apt to think, immutably,—as the alternation of day and night, summer and winter,—others contingently, as the motion of a body from its place, if pushed. . . ."

179. *Ibid.*, pp. 135–136: "The imposition of a name on any subject of contemplation, be it a material object, a phenomenon of nature, or a group of facts and relations, looked upon in a peculiar point of view, is an epoch in its history of great importance. It not only enables us readily to refer to it in conversation or in writing, without circumlocution, but, what is of more consequence, it gives it a recognized existence in our own minds, as a matter for separate and peculiar consideration . . . and . . . fits it to perform the office of a connecting link between all the subjects to which such information may refer."

180. Probably related to the family of Lord Clive (1725–1774), of whom there is a large statue in the center of Shrewsbury. A Miss Clive is mentioned in a letter dated December 15, 1824, written by Mrs. Josiah Wedgwood to her sister Fanny Allen: "I went as chaperon to the Drayton Assembly with Miss Clive, Susan Darwin, Charlotte and Fanny, with Joe and William and Edward Clive, but it was a bad and very thin ball and double the number of ladies to the gentlemen." Litchfield, *op. cit.*, Vol. 1, p. 163. Lord Clive was born near Market Drayton.

181. Darwin married Emma Wedgwood, January 29, 1839, i.e., about the time these pages were written.

182. Brougham, Henry Lord, *Dissertations on Subjects of Science Connected with Natural Theology: Being the Concluding Volume of the New Edition of Paley's Work*, 2 vols., Knight London, 1839, Vol. 1, p. 196: "Now connecting the two together [i.e., a particular action with a sign], whatever be the manner in which the sign is made, is Abstraction; but it is more, it is the very kind of Abstraction in which all language has its origin—the connecting the sign with the thing signified; for the sign is purely arbitrary in this case as much as in human language." In Darwin's copy, now in the Cambridge University Library, there is a vertical pencil line in the margin beside the text, and in Darwin's handwriting are the words, "don't understand." P. 197: ". . . a rational mind cannot be denied to the animals, however inferior in degree their faculties may be to our own."

183. Davy, *op. cit.*

184. *Ibid.*: "The desire of glory, of honour, of immortal fame and of constant knowledge, so usual in young persons of well-constituted minds, cannot I think be other than symptoms of the infinite and progressive nature of intellect—hopes, which as they cannot be gratified here, belong to a frame of mind suited to a nobler state of existence."

185. This incident is written out in greater detail in a note dated January 13, 1839. See *OUN* 31.

186. Sebright, John, *Observations upon the Instinct of Animals*, Gossling & Egley, London (pamphlet), 1836, 16 pp., pp. 15–16: "No one can suppose that nature has given to these several varieties of the same species such very different instinctive propensities, and that each of these breeds should possess those that are best fitted for the uses to which they are respectively applied. It seems more probable that these breeds having been long treated as they now are, and applied to the same uses, should have acquired habits by experience and instruction, which in course of time have become hereditary. . . . I am led to conclude, that by far the greater part of the propensities that are generally supposed to be instinctive, are not implanted in animals by nature, but that they are the result of long experience, acquired and accumulated through many generations, so as in the course of time to assume the character of instinct. How far these observations may apply to the human race I do not pretend to say; I cannot, however, but think that part of what is called national character may, in some degree, be influenced by what I have endeavoured to prove, namely that acquired habits become hereditary."

187. Gardiner, William, *The Music of Nature, or, an Attempt to Prove that What is Passionate and Pleasing in the Art of Singing, Speaking, and Performing*

upon Musical Instruments, is Derived from the Sounds of the Animated World, etc., Longman, London, 1832, p. 31.

188. Tooke, John Horne, *Epea Pteroenta, or the Diversions of Purley,* 2nd ed., Johnson's, London, Part I, 1798, Part II, 1805. (Copy in the Athenaeum Library.) See also *OUN* 5, and *OUN* 13.

189. See Beyle, Marie Henri (1783–1842), *The Lives of Haydn and Mozart, with Observations on Metastasia, and on the Present State of Music in France and Italy,* transl. from the French of L.A.C. Bombet [pseud.] with notes, by the author of the Sacred Melodies [W. Gardiner], 2nd ed., Murray, London, 1818, p. 115.

190. Wells, Algernon, *On Animal Instinct,* Colchester, 1834, 40 pp. (copy in the Congregational Memorial Hall Trust Library, Cricklewood, London, N.W. 2).

191. *Ibid.,* pp. 13–14: ". . . there is scarcely a faculty of mind or quality of character prevailing among men, but its type of resemblance may be found in some of the tribes of inferior creatures. Hence their actions and dispositions have ever furnished the moralist with those striking and instructive fables. . . . The industry of the bee and the ant—the cunning of the fox. . . ."

192. *Ibid.,* p. 15: "Instinct acts its part with unerring precision, without intelligently knowing what or why it does so. . . ." N.B.: Darwin indicates his disagreement with the first part of this sentence by saying, "no."

193. *Ibid.,* pp. 16–17: "Instinct is confined to narrow limits, but within them it never mistakes . . . it is observed,

"That the processes of reason and contrivance in men are capable of almost endless degrees of imperfection or improvement. . . . But instinct reaches its full perfection at once: and never afterwards receives, or admits of, any improvement. . . . the texture and shape of a bird's-nest, or of the cells and masses of honeycombs, are now what they ever were; and ever will be, without variation of improvement, or degeneracy."

194. *Ibid.,* pp. 18–19: "Besides which, many animals possess a strong imitative faculty . . . pure instinct is not an imitative faculty. . . . But imitation seems invariably associated with reason; is one of the most powerful laws by which it acts; and one of the most effective means of its acquisition and advancement."

195. *Ibid.,* pp. 18–19: "In those processes of instinct which are most difficult and surprising, it is impossible any part of the skill . . . should have been gained by imitation; especially in the case of numerous insect tribes, which never knew their parents. . . ."

196. *Ibid.,* p. 19: "Moreover, animals are capable of education: they may be, and often are, taught things that greatly surprise every beholder."

197. *Ibid.,* p. 19: "Now, instinct is neither knowledge gained by instruction, nor a faculty of being improved by instruction."

198. *Ibid.,* p. 20: ". . . the inferior creatures (inasmuch as they perform by far the greater number of their actions, especially in their wild, native state, by innate, blind instinct) may be properly denominated *creatures of instinct;* although . . . they are not bound down to instinct as their only means of knowledge and action. Just as, on the other hand, man is properly denominated *the creature of reason* . . . some of his actions are instinctive; performed especially in infancy. . . ." The expression *"creatures of habit"* does not appear on these pages in Well's text.

199. *Ibid.,* p. 22: "Distinct notice should be taken of the curiously perfect adaptation of the instincts of animals to their senses and bodily structure; and of both to those scenes or portion of the external world in which it is designed they should dwell. . . ."

200. Darwin has reference to the "Woodpecker of the plains," *Colaptes campestris,* which he observed on the northern bank of the Plata, in Banda Oriental, South America, and which rarely visited trees. See Darwin, Charles "Note on the Habits of the Pampas Woodpecker *(Colaptes campetris)*," *Proceedings of the Zoological Society of London,* 1870, pp. 705–706.

201. Wells, *op. cit.,* pp. 28–29: "But to observe a bee, at the distance of a mile or more from its hive, busy among the flowers, without the least anxiety lest it should be lost amidst its mazy flights; and, when loaded, wing its direct way to the hive, without thought, and yet without error, is to us amazing."

202. *Ibid.*, pp. 29–30: "No faculty we possess [as the carrier-pigeon] helps us to any analogy by which to enable us to form any notion of such a power. It is intuition—it is inspiration—it is something we do not possess, and cannot conceive of. . . . It is one of those wonders with which the works of God abound. . . ."

203. Emma, Darwin's wife.

204. V. E., i.e., Vide Notebook E. Darwin inadvertently entered on p. 125 of Notebook E the following: "Uncovering the canine teeth or sneering, has no more relation to our present wants or structure, than the muscles of the ears to our hearing powers. E. frowns prodigiously when drinking very cold water, frowns connected with pain as well as intense thought." The E. in the quotation is probably Emma, but, as de Beer, Rowlands, and Skramovsky (1967, p. 173) indicate, it could also be Erasmus, his brother. (de Beer, Gavin, M. J. Rowlands, and B. M. Skramovsky, eds., "Darwin's Notebooks on Transmutation of Species, Part VI. Pages Excised by Darwin," *Bulletin of the British Museum* (*Natural History*) *Historical Series*, 3 (5) :129–176, 1967.)

205. The lower half of the page has been excised, affecting pages 73 and 74. Because the cut was a curving line, a word or two is missing from the middle of the last remaining line on p. 74.

206. Henslow, John Stevens, Professor of Botany, Cambridge.

207. Necker, Albertine Adrienne, *Progressive Education; or, Considerations on the Course of Life,* transl. from the French of Mme. Necker de Saussure, 3 vols., Longman, etc., London, 1839, Vol. 1, p. 4: "St. Paul tells us that we have two laws within us* (*Romans, vii, 23) ; and our inward feelings, our experience, our reason, all confirm this declaration. A blind instinct, necessary perhaps to the physical order of things, impels us to seek after pleasure, and thus favours the developement of our faculties. . . ." P. 40: ". . . amongst all these philosophers [astronomers, etc.,] there is not one father who has taken the trouble to note down the progress of his own child." Perhaps it was due to this suggestion of Mme. Necker that Darwin did observe and record the progress of his oldest child.

208. The lower half of p. 76, beginning with "April 3d, 183[9]," was excised.

209. The lower half of p. 76 (verso 75) , beginning with "have been carried," was excised.

210. "Turks" difficult to decipher.

211. Mackintosh, *op. cit.*, 1837, pp. 96–97.

212. Bell, Charles, *The Hand, Its Mechanism and Vital Endowments as Evincing Design* (Bridgewater Treatises on the Power, Wisdom, and Goodness of God, as manifested in the Creation. Treatise IV) , Pickering, London, 1833, p. 191.

213. See Lamarck, *op. cit.,* in section "Of Reason. And Its Comparison with Instinct,"

214. See Lamarck, *ibid.,* in section, "Of the Emotions of the Inner Feeling "

215. See Lamarck, *ibid.,* in section "Of the Origin of the Propensity Towards Repeating the Same Actions, and Also of Instinct in Animals": "Who can now deny that the powers of habits over actions is inversely proportional to the intelligence of the individual, and to the development of his faculty of thinking, reflecting, combining his ideas, and varying his actions?" See also *M* 108.

216. See Lamarck, *ibid.,* in section "Of Instinct in Animals": ". . . the habit of using any organ or any part of the body for the satisfaction of constantly recurring needs, gives the subtle fluid so great a readiness for moving towards that organ where it is so often required, that the habit becomes inherent in the nature of the individual.

"Now the needs of animals with a nervous system vary in proportion to their organisation, and are as follows:

1. The need for taking some sort of food;
2. The need for sexual fertilisation, which is prompted in them by certain sensations;
3. The need for avoiding pain;
4. The need for seeking pleasure or well-being.

"For the satisfaction of these needs they acquire various kinds of habits, which become transformed in them into so many propensities; these propensities they cannot resist nor change of their own accord. Hence the origin of their habitual actions and special inclinations, which have received the name of instinct.*" (Footnote: "*Just as all animals do not possess the faculty of will, so too instinct is not a property of all existing animals; for those which have no nervous system have no inner feeling, and cannot therefore have any instinct for their actions. These imperfect animals are entirely passive, do nothing of their own accord, feel no needs, and are provided for by nature in everything just as in the case of plants. Now since their parts are irritable, nature causes them to carry out movements, which we call action.")

"This propensity of animals to the preservation of habits, and to the repetition of the resulting actions when once it has been acquired, is propagated to succeeding individuals by reproduction so as to preserve the new type of organisation and arrangement of the parts; thus the same propensity exists in new individuals, before they have even begun to exert it.

"Hence it is that the same habits and instinct are handed on from generation to generation in the various species or races of animals, without any notable variation so long as no alteration occurs in their environment."

217. The "something" is "Natural Selection."

218. Mackintosh, *op. cit.*, pp. 152–153: "Reason, as reason, can never be a motive to action. It is only when we superadd to such a being [viz., one capable of reason but incapable of pain or pleasure] sensibility, or the capacity of emotion or sentiment (or what in corporeal cases is called sensation), of desire and aversion, that we introduce him into the world of action." See also n. 88, *OUN* 51.

219. This reference has not been traced.

220. Bennett, George, *Wanderings in New South Wales, Batavia, Pedir Coast, Singapore, and China; being the Journal of a Naturalist in those Countries, during 1832, 1833, and 1834*, 2 vols., Bentley, London, 1834, Vol. 1, p. 234.

221. Gardiner, *op. cit.*, p. 199: "Dogs in a state of nature never bark; they simply whine, howl, and growl. . . ." See also *Zoonomia*, pp. 154–155, for a discussion of dogs of Juan Fernandes and Guinea that do not bark.

222. Bennett, *op. cit.*, Vol. 2, p. 153: "He [a male Ungka gibbon (*Hylobates syndactyla*) taken on board ship] could not endure disappointment, and, like the human species, was always better pleased when he had his own way; when refused or disappointed at anything, he would display the freaks of temper of a spoiled child; lie on the deck, roll about, throw his arms and legs . . . dash everything aside that might be within his reach . . . he reminded me of that pest to society, a spoiled child. . . ." P. 154 (footnote): "The account of the orang-utan, given by Dr. Abel, in the *Narrative of a Journey in the Interior of China*, accords with the habits of this animal, and the comparison is very interesting."

223. Abel, Clarke, *Narrative of a Journey in the Interior of China, and of a Voyage to and from that Country, in the Years 1816 and 1817*, etc., Longman, London, 1818, p. 326: "If defeated again by my suddenly jerking the rope, he [the orang] would at first seem quite in dispair, relinquish his effort, and run about the rigging screaming violently." P. 328: "If repeatedly refused an orange when he attempted to take it, he would shriek violently and swing furiously about the ropes; then return and endeavor to obtain it; if again refused he would roll for some time like an angry child upon the deck uttering the most piercing scream; and then suddenly starting up, rushing furiously over the side of the ship and disappear."

224. Cowper, William (1731–1800) see *Works, Comprising His Poems, Correspondence and Translations*, Robert Southey, ed., 15 vols., Baldwin & Cradock, London, 1835–1837.

225. The reference to Mr. Hamilton on vital laws could not be traced in the Athenaeum Library, Pall Mall, London.

226. Upper portion of pp. 99 and 100 excised.

227. Hume, *op. cit.*, Vol. 1, Part III, p. 232, "Of the Reason of Animals," and Vol. 4, Section 9, p. 121, "Of the Reason of Animals." The edition Darwin used has not been traced.

228. *Ibid.*, Vol. 2, p. 419, "Dialogues Concerning Natural Religion."

229. *Ibid.*, Vol. 4, Section 4, p. 32, "Sceptical Doubts Concerning the Operations

of the Understanding," and Section 5, p. 49, "Sceptical Solution of those Doubts."

230. *Ibid.*, Vol. 4, pp. 435–510, "The Natural History of Religion."

231. Undoubtedly a personal communication.

232. Akerly, Samuel, "Observations on the Language of Signs, Read Before the New York Lyceum of Natural History, on the 23d June, 1823," *American Journal of Science and Arts*, 8 (2) :348–358, 1824, p. 351.

233. Parsons, James, "Human Physiognomy Explain'd: in the Crounian Lectures on Muscular Motion," Lecture 1, pp. 1–31; Lecture 2, pp. 32–82, plus Index, 4 pp., *Philosophical Transactions*, 44 (Part 1) :1–82, 1746.

234. Brun. See Lavater, Gaspard, *L'Art de Connaître les Hommes par la Physionomie*, 10 vols., Paris, 1820, Vol. 9, p. 268, "Conférence sur l'Expression." Ref.: Darwin, Charles, *The Expression of the Emotions in Man and Animals*, Murray, London, 1872, n. 6, p. 3. See also *ibid.*, pp. 4, 247, and 287, for discussions of Le Brun's description of the expression of fright, of anger, and of astonishment.

235. In *Zoonomia*, p. 155, is the following similar passage: ". . . the singing of birds, like human music, is an artificial language rather than a natural expression of passion.

"Our music, like our language, is perhaps entirely constituted of artificial tones, which by habit suggest certain agreeable passions."

236. Cresselly, Pembrokeshire, home of John Bartlett Allen (1733–1803), father of Elizabeth (1764–1846), the wife of Josiah Wedgwood (of Maer) and mother of Emma, whom Darwin married.

237. This reference has not been traced, but probably W. Ogilby, author of various papers on gibbons, lemurs, and monkeys reported in the *Proceedings of the Zoological Society of London* in 1837 and 1838. Perhaps the reference is to a remark made by Ogilby at a meeting attended by Darwin.

238. "Descent 1838" and seven or eight illegible words written in faint pencil across the page.

239. Probably Eyton, Thomas Campbell (1809–1880), companion of Darwin during his youth.

240. Ernest Wedgwood (1838–1898), son of Hensleigh and Frances Wedgwood.

241. Frances Julia (Snow) Wedgwood (1833–1915), sister of Ernest.

242. Katherine Euphemia (Effie) Wedgwood (1839–?), another sister.

243. Goldsmith, Oliver, *The Miscellaneous Works of Oliver Goldsmith, M.B.*, 4 vols., Murray, London, 1837, Vol. 4, No. XV, "Spenser's Faerie Queene."

244. Smith, Adam, *The Theory of Moral Sentiments; or, An Essay Towards an Analysis of the Principles by which Men Naturally judge concerning the Conduct and Character, first of their Neighbours, and afterwards of themselves, to which is added, a Dissertation on the Origin of Languages.* 10th ed., 2 vols., Cadell & Davies, London, 1804. (Copy in The Athenaeum Library.)

245. Hume, David, *Essays and Treatises on Several Subjects*, 2 vols., Bell, etc., Edinburgh, 1817, Vol. 2, pp. 170–207. "A Dissertation on the Passions."

246. Hartley, David, *Observations on Man, His Frame, His Duty, and His Expectations: In Two Parts: To which are now first added, Prayers, and Religious Meditations. To the First Part are Prefixed, A Sketch of the Life and Character, and a Portrait, of the Author*, 5th ed., 3 vols., Crattwell, Bath, 1810. (Copy in The Athenaeum Library.)

247. Probably Thomas Brown, M.D. (1778–1820), Professor of Moral Philosophy at the University of Edinburgh, author of *Observations on the Zoonomia of Erasmus Darwin, M.D.*, Mundell, Edinburgh, 1798; *Lectures on the Philosophy of the Human Mind*, 4 vols., Tait, Edinburgh, 1820; *Sketch of a System of the Philosophy of the Human Mind*, Part first, *Comprehending the Physiology of the Mind*, Bell & Bradfute, Edinburgh, 1820.

N Notebook

Commentary by Howard E. Gruber

"Experience shows the problem of the mind cannot be solved by attacking the citadel itself.—the mind is function of body." [*N* 5]

pp. 1–5 *Moment of truth.* The N notebook is simply a continuation of the M notebook, and yet it comes at a significant moment in Darwin's thought. On September 28, 1838 as recorded in the D notebook, provoked by a suggestive passage in Malthus' *Essay on Population,* Darwin had his first clear insight into the principle of evolution through natural selection. Since the N notebook takes up only a few days later, we have an opportunity to discover whether this event had any immediately transforming effect on Darwin's ruminations about man and mind.

Motives in conflict. The entry for October 2 is simply a few facts about bird behavior and emotional expression, garnered from reading, from talk with his father, and from his own observations.

The entry for October 3 deals with the origin of abstractions such as conscience and the idea of God. Darwin begins his argument by noticing that animals can undergo conflicts of motives— like a dog stimulated at one and the same time to chase a rabbit and to scratch for fleas. Since the animal will have to choose between alternative courses of action, it will have to slight one motive, or disobey "a wish which was part of his system." If the creature remembers and thinks about what it has done, it will be in a state quite analogous to a "troubled conscience." Although Darwin introduces the point with an example of competing *non*-social motives (to chase or to scratch), he soon makes it clear that he really has in mind those questions which are matters of conscience for man: "Therefore I say grant reason to any animal with social & sexual instincts & yet with passion he *must* have conscience." (*N* 2–3)

Having satisfied himself that "moral sense" must be a universal attribute of man, Darwin has now to counter the objection that it is not universal, since it varies from one society to another. Not at all, says Darwin, the fact that the moral sense, like all other attributes, varies so as to adapt itself to different circumstances demonstrates its real existence.

The passage closes with a brief remark suggesting that the idea of God may also arise as a consequence of certain general features of the human condition. It is too bad that we cannot make out the exact sense of this important sentence. There are one or two indistinct words, and only the general sense is clear.

The second entry for October 3 is fascinating. He begins by recording an instance of unlearned, beautifully coordinated visual-motor behavior in a newly hatched chick. He points out that the same behavior in man would require experience. This is a puzzle.

Such puzzles cannot be solved by a direct philosophical ("metaphysical") attack. Just as the science of astronomy could be perfected only when Newton linked it with the laws of mechanics, the problem of mind requires "some stable foundation." Exactly what that foundation might be he does not say, but, judging from his own efforts, he seems to mean a scientific psychology based on an evolutionary theory.

> "Hope is the expectant eye looking to distant object, brightened & moistened by emotion,—why does emotion make tears fall?" [N 9]

pp. 6–10   *Expression of emotion.* There is nothing very new in this passage. Darwin notes several analogies between emotional expression in man and animals. He gives a few examples of the law of antithesis, discussed above, and adds: "the hypothesis of opposite muscles will want much confirmation." (N 8) He touches on the expression of a variety of emotions, always looking for a description that will clarify the exact musculature involved in facial expression, the adaptive significance of the acts, and the resemblances between man and other animals.

The remarks about laughter show a trace of interest in inappropriate emotional expressions when he speaks of *"senseless* grin of passion," but he is making only a limited point, since the expression is a "symbol of readiness" to bite. This is definitely not the point of view we find in *Expression,* where he borrows from Herbert Spencer and others the hydraulic, pre-Freudian notion that emotion may spill over from its appropriate channels of expression.*

---

* *Expression,* Chapter 8: "Joy, High Spirits, Love, Tender Feelings, Devotion."

"After the desire of food, the most powerful and general of our desires is the passion between the sexes, taken
in an enlarged sense. Of the happiness spread over human
life by this passion very few are unconscious." [Thomas
Malthus]*

pp. 10–
13

Malthus and Darwin. Here we have Darwin's first reference to
Malthus in the M and N notebooks. It is obvious that Darwin is
interested in the origins of morals. Malthus' main argument is a
plea for the limitation of family size, so that the population will
not outstrip the available food supply. The means by which Malthus hoped to accomplish this goal was not contraception, which
he considered to be improper, but sexual continence and late marriage, especially on the part of the poor.

Now that is the dour Rev. Malthus whom we all know, and
whom Darwin knew too—but Darwin picked up something else
from Malthus, something he surely wanted to find, a warmer note,
extolling the virtues of sexual love when properly regulated and
directed. And so, Darwin writes, see Malthus "on passions of mankind, as being really useful to them: this must be studied before
my view of origin of evil passions." (*N* 10–11) From this sentence
it is clear that my earlier guess was correct—the "evil passions"
Darwin had in mind on page *M* 123 have more to do with sex
than with anger.

A number of previously mentioned points are briefly repeated
in this passage: the comparison between visual development of
chickens and men, the insistence on "necessary notions," similar
to abstract ideas, in animals, the enthusiasm for Comte's conception of three stages in the history of each science. Darwin adds,
"Zoology itself is now purely theological." (*N* 12)

Can plants think? At this point Darwin is conceding nothing in
his search for continuities between man and all other living
creatures: "have plants any notion of cause and effect. They have
habitual action which depends on such confidence—when does
such notion commence?" (*N* 13)

The possibility that plants think may strike the reader as plain
silly. For Darwin, entertaining the idea was just pushing a bold
idea all the way. Years later he made the idea a lot more plausible
with his works on the behavior of plants—climbing plants, twin-
ing plants, and insectivorous plants.

---

* Malthus, *op. cit.*, Vol. 2, p. 154.

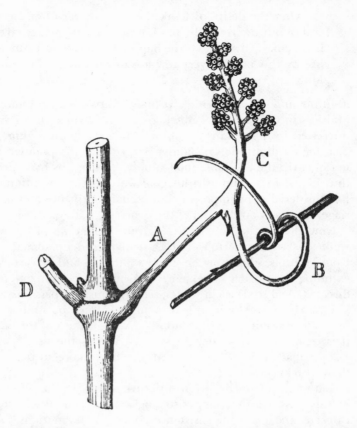

"Flower-stalk of the vine. A. Common peduncle. C. Sub-peduncle, bearing the flower-buds. B. Flower tendril, with a scale at its base. D. Petiole of the opposite leaf." Adaptive behavior in plants: the tendril of a vine has coiled around a neighboring branch to support the weight of new growth.

> "All Science is reason acting, systematizing, on principles, which even animals practically know . . . an ass knows one side of triangle shorter than two." [*N* 14]

pp. 13–
18

Science and common sense. We hear a complex medley of themes, each expressed in a swift allusion. It is hard to make sense of the passage as a whole, but I think it has one central idea. Practical knowledge, intuition, and art precede abstract science in the development of human thought. In the same manner, a history of conflicting impulses and spontaneous actions precedes the development of value judgments and moral sense.

In the field of practical mechanical affairs, an intuitive grasp of geometry is implied in adaptive behavior; our abstract ideas of

necessary truth "follow as consequences . . . to propositions which
are the result of our senses, or our *experience*." (*N* 16)

This empirical approach to the basis of geometrical propositions
has, of course, its own complex history. Darwin's empirical attitude toward propositions—that they are based on the experience
of the individual or of the species, and that they are not necessary
ideas in any absolute sense—is a reflection of a very general current in nineteenth-century thought.

Likewise, in matters of desire, when the organization of the
brain has evolved sufficiently to permit adequate memory, "comparison of sensations would first take place whether to pursue immediate inclination or some future pleasure,—hence judgment,
which is part of reason." (*N* 18)

The fragmentary nature of this passage is, I believe, due mainly
to the interweaving of these themes. One ought to note, however,
that three days earlier, October 5, he began working on a very
demanding task, and may have been giving less attention to the
subjects of the M and N notebooks. His *Journal* contains the following entry: "October 5th Began Coral Paper requires much
reading."

His reference to Whewell on the relation between science and
common sense probably does not mean that he was really reading
Whewell at the time. Whewell had been one of Darwin's teachers
at Cambridge years before, and the latter probably read the book
when it appeared in 1837. He refers to it earlier in one of the
transmutation notebooks. He could remember where something
was on the page,* and might simply have looked up the reference
while writing down his own thoughts.

The paragraph on blushing grew into a chapter in *Expression of
Emotions*. There he stressed the universal occurrence of blushing
among all races of man, its non-occurrence in other animals, and
its unlearned character. The explanation of blushing he developed
in 1872 is not yet suggested in the present paragraph, but see *N* 52.

"The taste for recurring sounds in Harmony common
to whole kingdom of nature." [*N* 18]

pp. 18– *Higher and higher mental powers.* Darwin clearly has a program
24   to derive higher mental powers, such as language and abstract

---

* In his reply to Francis Galton's questionnaire on imagery, a landmark in the
history of psychology, Darwin describes himself as having fairly strong visual
imagery. Regarding images of printed matter he wrote, "I cannot remember
a single sentence, but I remember the place of the sentence and the kind of
type." (*LL* 3, 239)

thought, from simpler processes. He wants to show that, given the simpler ones, the higher powers must necessarily evolve. Given memory, the comparison of things previously perceived is bound to occur, and this is the beginning of reasoning. Given certain expressive sounds—crying, yawning, laughing—produced involuntarily and always in association with certain events, language can emerge from the "necessary connexion between things & voice." (*N 20*)

There are two distinct trends in Darwin's remarks about language here, emotional expression and the association of ideas. In the *Descent of Man* these are linked in an interesting way. Patterns of sound during courtship and other emotional expressions provide a basis for sexual selection and the consequent improvement of the organs of vocalization. At the same time, some of these expressive sounds come to be associated with the acts in which they occur, so that some primordial "words" arise almost directly as emotional expressions. On this basis, and with the improved brain and vocal organs resulting from evolution up to that point, further associations of sounds and things can develop. The same ideas appear later in the *Descent of Man* in various passages on language, music, and speech.

The remark on prayer seems to suggest that, although religious feelings are very remote from the simplest thoughts, a prayer may be thought of as an "eloquent request"—and the ability to make such a request emerges from a growing power to think about absent things. Judging from the next sentence, Darwin may have been drawing an analogy between the continuum from simple desire to eloquent prayer and another series—from song to poem. From the corresponding passage in *Descent* we may suppose that Darwin was groping for the idea that prayer represents desire infused with poetry:

"The feeling of religious devotion is a highly complex one, consisting of love, complete submission to an exalted and mysterious superior, a strong sense of dependence, fear, reverence, gratitude, hope for the future, and perhaps other elements. No being could experience so complex an emotion until advanced in his intellectual and moral faculties to at least a moderately high level. Nevertheless, we see some distant approach to this state of mind in the deep love of a dog for his master, associated with complete submission, some fear, and perhaps other feelings." (*Descent*, 95–96)

Shrugging the shoulders. The passage in these notes resembles, but is not as clear as, the eventual treatment of the subject in *Expression*. In the notes he thinks "shrugging connected with many

emotions." In *Expression* it is an expression of helplessness and
impotence used when someone "wishes to show that he cannot do
something or prevent something being done." In *Expression* we
learn that shrugging is found all over the world in all races of
men, and that it is unlearned, since Laura Bridgman, who was
blind and deaf and could not have learned to shrug by imitation,
shrugged like the rest of us. Both in the notes and the book Darwin treats shrugging as another case of the law of antithesis—the
actions involved are opposed in kind to what we must do to act
defiantly. (*Expression*, Chapter 2)

> "the possibility of poets describing gentle things in
> gentle language, & vice versa,—almost proves that at ear-
> liest times there must have been intimate connection be-
> tween *sound* & language." [*N* 31]

pp. 25–
33

The origin of taste. In Darwin's view, aesthetic taste is acquired,
both in the history of the individual and in the history of the
species. Some idea of beauty is universal, but the particular objects to which we attach it depend on our inheritance from our
immediate ancestors: "a mountaineer born out of country yet
would love mountains"—he would not need personal experience
with mountains to acquire a taste for them. (*N* 26)

But Darwin does not seem to mean that there is an entirely arbitrary connection between experience and our ideas of beauty.
He cites, a bit vaguely, two counter-examples. There is something
intrinsically ugly about "great masses of rock"—which, because of
the connection, makes new buildings look ugly, but not old ones
like Windsor Castle. In language there are intrinsically gentle
sounds, and other sounds for other feelings, which accounts for
some of the expressive power of poetry.

In two ways, then, Darwin avoids a radical empiricism: The
relevant experience need not occur in the individual life history;
and the development of aesthetic taste depends, at least in part,
on certain intrinsic properties of the sensuous experience.

In his quotation from the *Discourses on Art* of Sir Joshua Reynolds, his own point of view is clear. There is both a universal and
a particular aspect to the expression of feeling: " 'The general
idea of showing respect is by making yourself less, but the manner . . . is matter of custom.' " (*N* 32)

On October 25 Darwin went to Windsor for two days' rest.
Characteristically, he found something there to nourish each of a
number of trains of thought. In the E notebook he records some

observations on the varied markings of fallow deer in Windsor Park, suggesting to him that variation under domesticity is similar to that in the wild state. And he saw a hybrid between a silver fish and a gold fish—hybridization was one of his favorite subjects. In the Castle itself and in its landscaping he found, as we have seen, something to think about and admire.

"Why should you or I speak of variation as having been ordained and guided, more than does an astronomer, in discussing the fall of a meteoric stone? He would simply say that it was drawn to our earth by the attraction of gravity, having been displaced in its course by the action of some quite unknown laws."*

pp. 33–36   *Creation versus law.* Darwin records another dream. As before, he is interested not in the content of the dream but in its most general psychological characteristics. He remarks on the disparity between his ability to read French when awake and in a dream, and on the disconnected character of dream fragments.

After a brief comment on boyhood morality, he goes on to register a protest against a contradiction in contemporary thought: it is accepted that the whole Newtonian universe of the astronomers, vast and awesome as it may be, operates according to natural law;

N Notebook, p. 36. *Courtesy of Cambridge University Library.*

---

* Letter, Charles Darwin to Charles Lyell, August 21, 1861. *ML* 1, 194.

but the much more limited universe of living things is deemed so wonderful as to require the idea of a special intelligence, "the artificer," who designs and creates each living thing. Darwin adds a reflection on human psychology: if we admired the universe as much as we did "the wonderful structure of a beetle," we would apply the argument from Design to the former as much as to the latter, and not accept the operation of either according to natural law.

"child . . . when refusing food, turn his head first to one side & then to other, & hence rotatory movement negation." [*N* 37]

PP. 37–40   *A family visit.* On November 11 he had successfully proposed marriage to Emma Wedgwood, and spent the next week in visits between the two family homes, returning to London on November 20, 1838, the date of this passage.

Not surprisingly, the first remark is an observation he has made about the behavior of a baby nephew. It concerns the spontaneous and natural emergence of a language of gestures before speech. This is followed by a note on something his sister has said about the baby's ability to recognize facial expressions of emotion.

He then notes two medical examples, presumably drawn from conversation with his father, of the failure of voluntary control of behavior. In both cases, attempting to prevent an involuntary act simply draws the person's attention to the part of the body involved and increases the likelihood of the act. This is yet another example of Darwin's consistent interest in the dethronement of reason and conscious will as the sole governors of human behavior.

From a conversation with his brother-in-law he has picked up a snippet about onomatopoeia in Scottish poetry, which Darwin thinks can be used to support his view that language originated in emotional and imitative expression.

The passage ends with a brief allusion to the development of complex habits and instincts in creatures of "very limited reasoning powers." While the exact train of thought is hard to follow, it is amusing to see Darwin link up in one sentence the Birgos crab which he had observed during the voyage of the *Beagle* with the "children and old people" he has just been visiting.

"The other principle of those children which *chance* produced with strong arms, outliving the weaker ones, may be applicable to the formation of instincts, independently of habits." [N 42]

pp. 41–
44 (See
Beagle
Diary,
367)

Kissing. In his *Journal* for 1838 Darwin wrote, "Wasted entirely the last week of November." A good deal of time went into house-hunting in London in preparation for the forthcoming marriage.*

There is something very personal in his way of discussing the connection between salivation, sexual desire, and a suggestion as to the evolution of kissing. He speaks almost in the first person of "ones tendency to kiss, & almost bite, that which one sexually loves." (N 41) At the time he wrote this paragraph he seems to have believed that kissing was universal in humans; by the time he wrote the *Expression of Emotions* he was better informed about the variety of ways of expressing affection—he gives a brief sketch of this diversity and stresses only the universality of affectionate behavior in mammals. (*Expression*, 222–23) In the 1838 notes he is quite pleased with himself for having created a theory which renders intelligible, through the tortuous path of evolution, the fact that in both humans and dogs "sexual desire makes saliva to flow." (N 41)

Natural selection in man. This is the first passage in the M and N notebooks in which the principle of evolution through natural selection is clearly stated and applied to man. Darwin sees it as one of *two* principles which can account for evolutionary change, the other principle being the direct inheritance of acquired characteristics. In the present passage he indicates that he has no way of deciding upon the relative importance of the two principles, and that both may be important.

Hybridization and behavior. He goes on to describe several examples of canines in which cross-breeding produces offspring that are intermediate both in structure and in behavior, a fact which he wants to relate to the more general question of the relation of mind and body: in hybrids, "can we deny that brain would be intermediate like rest of body? Can we deny relation of mind and brain." (N 43)

This passage, then, makes reasonably explicit *three* possible sources of evolutionary variation: inheritance of acquired characteristics, chance variation, and hybridization. It is noteworthy that this first explicit reference to natural selection as applied to man comes at just this moment. Although Darwin had seen the main point on September 28, 1838, it was not until about two months later that he re-stated the theory of evolution though natural

---

* *Emma Darwin: A Century of Family Letters.* Letter from Charles to Emma, November 30, 1838, Vol. 2.

selection in a brief and logical form. Now he can see with new
clarity that he does not have to rely on the inheritance of acquired
characteristics as the mechanism of evolutionary change. Since the
development of this latter principle was originally the main purpose of the M and N notebooks, we may now expect to see some
significant changes in the character of these notes.

> "In my theory there is no absolute tendency to pro-
> gression, excepting from favorable circumstances!" [*N* 47]

pp. 45–
49

The idea of progress. Darwin walked a tightrope on the question
of progress. He thought it to be a widespread phenomenon—in a
broad evolutionary sense, in the history of man, and in the development of the individual. Widespread, but neither ubiquitous
nor inevitable—dependent, rather, on the complex interplay of
natural forces. Given this delicate balance, one can sometimes
catch him vacillating, or seeming to.

In the present passage he asserts that there is no absolute law
of progress, and then goes on in the next sentence to insist "that
it requires a far higher & far more complicated organization to
learn Greek, than to have it handed down as an instinct." (*N* 48)

Those remarks, written on November 28, 1838, ought to be contrasted with an earlier passage in which he criticizes the anthropo-
centric view that some mental faculties are higher than others.
(*B* 74)

The reader can try for himself to catch the flavor of Darwin's
thoughts by trying to reconstruct the connections between ideas in
this passage full of bewildering jumps. Some of my guesses are as
follows: He has the notion of progress from lower faculties to
higher ones in mind when he begins to write. This leads him to
search afresh for examples of continuities—physical pain and
mental grief, "old hereditary ideas" and youthful "acquirement of
new ideas," a dog's loyalty and a man's conscience. At this point
he feels ready to try for a generalization or two—the two paragraphs on progress concluding, "Instinct is a modification of bodily
structure . . . to obtain a certain end; & intellect is a modification
of instinct—an unfolding & generalizing of the means by which an
instinct is transmitted." (*N* 48)

In the next paragraph he feels a need for some philosophical
justification of his method of reasoning. The thought is com-
pressed but he seems to be saying that it is legitimate in science to
look at extreme cases, in which the effect we want to study is mag-
nified—thus, to understand the evolution of mind in lower ani-

mals, we may look at the "frontier instance," man. But, he warns himself parenthetically, although you may think of man as the "highest," remember that he is only an extreme case, he is not a deity, not a causal agent. Causes are to be found in nature, not in man or God.

This brings him back to one of his favorite subjects, free will. We may experience ourselves as causal agents having free will, but our desires and purposes do not arise out of some special endowment, they result from nothing but the natural laws of thought, something we could see if we only stand outside ourselves. But this we cannot do, hence "on my view of free will, no one could discover he had not it." (*N* 49)

"animals, not being such thinking people, do not blush."
[*N* 52]

pp. 50–55   *Mind over body.* Of all the pages in these notebooks, this passage most plainly invites consideration on two levels—the manifest scientific content which will be dealt with here, and the thinly veiled latent content which suggests Darwin's sexual anxieties and anticipations, aroused by thoughts of his impending marriage, and more especially by Emma's three-week stay in London, December 6–21.

In this passage he is writing, as often before, about the relation between mind and body.* But there is a reversal. He is talking about what we would now call *psychosomatic* or, better, *psychogenic* phenomena—the way in which mental events influence bodily ones.

Blushing. In this connection, he advances his theory of blushing, very much as it eventually appeared in the *Expression of Emotions.* Blushing occurs when blood rushes to some part of the body: the rush of blood is caused by the person's giving special attention to the part of the body in question; he does so because he thinks someone else is looking at it. From this it follows that we blush only in exposed parts of the body, and also that blushing should be occasioned by social contacts with the other sex, "because each sex thinks more of what another thinks of him, than of any one of his own sex." (*N* 52)

Having developed the point in connection with blushing, Darwin applies the general line of thought about mind-body relation-

---

* By this time we can be confident that for Darwin this means the relation between the brain and other organs of the body.

ships·in quick succession to male erections, toothache, and cancer.

Since his theory of blushing is founded on one person's aware-ness of the thought of another, he turns his attention to that sub-ject and gives two examples—a mother identifying with her son, and an unspecified person (whose sex changes from one sentence to the next) blushing with shame at the thought that some con-cealed act will be discovered by another.

All of this means that blushing requires a person conscious of himself and conscious of others, conscious of his own past and of others' thought about it. Animals do not blush.

"But why does joy, & *other emotions* make grown up people cry What is emotion?" [*N* 57]

pp. 56–
62

Emotion. These are desultory notes kept in the month before and the month after his marriage—some new reading and some new thoughts, but not very much. From his *Journal* it is clear that he did not do much work in the month preceding the marriage. About a week after it, he began studying German, and took up once more his work on coral reefs. By the time another month had passed, he was working hard again at various subjects, but the M and N notebooks never regained their earlier momentum.

Up to this point the notes have said a great deal about the *expression* of emotion, but very little about its actual nature. Here, after a few notes on expression, regarding fear and sobbing, Dar-win suddenly asks, as though it is a new question, "What is emo-tion?" He does not pursue the question at all vigorously, but ten days later the question recurs, and we can get some glimmering of how he might have pursued it: "What passes in a man's mind when he says he loves a person . . ."—he thinks it may be some combination of an image of the loved one and "blind feeling, something like sexual feeling." (*N* 59) Another question flits by—what is the interrelation *among* emotions? Is one emotion in-fluenced by other emotions?

Space, time, and causality. Darwin is interested in the origin of all ideas, but the idea of *cause* is particularly important to him, because of its intimate connection with the idea of *deity*. Ideas of cause are closely bound up with ideas of space and time.* A week before his marriage, we find him reading Sir John Herschel, his favorite author on the subject of scientific method. The brief note

---

*On the origin of ideas of space, time, and causality, see Jean Piaget, *The Child's Construction of Reality* (London: Routledge and Kegan Paul, 1955).

tells us that he is still trying to avoid the extreme empirical view of causality: the idea of cause is more than a simple induction drawn from observations of temporal succession—"succession of night & day does not give notion of cause." (N 60) The idea of necessary connection goes well beyond mere succession.

In the next paragraph we see a suggestion that he is applying an interesting psychological method to this philosophical question. If certain ideas, such as space, time, and causality, are logically interconnected, then we should expect to find correlations among them in their growth and decline in human thought. He gives a brief account of some material on senility from his father's medical experience, bearing on this point, and then raises a question about children.*

"that is my new part of the view.—let the proof of hereditariness in habits be considered as grand step if it can be generalized." [N 63]

pp. 62–68   *Unconsciousness and hereditary habits.* It is over four months since Darwin's first clear and great insight into the idea of evolution through natural selection, over two months since he stated his theory logically and succinctly as three principles, and also two months since he applied the theory explicitly to man. Nevertheless, he has not abandoned his earlier position that the inheritance of acquired characteristics can account for a large part of evolutionary change.

The present passage re-states an idea now familiar to us—habits acquired during the lifetime of the individual can induce structural changes, which can become hereditary, so that habits are passed on through the reproductive system from one generation to the next.

But he sees that he must think a little more deeply about the nature of such habits: They are distinct from conscious memories. There are numerous prominent examples of behavior controlled by habit without the individual's being aware of the sources of his actions. He cites habits formed in infancy which endure long after conscious memory for the original events has disappeared; a case of involuntary action on his father's part suggestive of automatic writing; and he deals with unconscious "mental impulses"—

---

* For a classic example of the use of psychopathological, anthropological, and developmental evidences in the study of questions of this kind, see Heinz Werner, *Comparative Psychology of Mental Development*, revised edition (Chicago: Follet, 1948) .

necessary to explain hereditary patterns of animal behavior. Although he realizes that he needs some such motivational concept as unconscious mental impulse, he is reluctant to be drawn into using Lamarck's notion that behavior is directly controlled by the will—this would run counter to his whole stress on involuntary action—and he sees a way out: "an animal may so far be said to *will* to perform an instinct that it is uncomfortable if it does not do it." (*N* 67) This solution to the problem can be seen either as question-begging or as a typical example of Darwin's sense of all living systems as delicate balances among many forces.

Emotions, language, children, and taste. These thoughts are interspersed with a number of other themes: the continuity between animal and human aesthetic tastes, children's ability to express themselves before they can speak, the onomatopoetic theory of the origin of language, and a note on a vestigial habit: children's pleasure in games of hiding, seen as related to instinctual concealment in animals—for example, young pigs.

In his concern for the nature of emotions, he detects an important paradox. If emotions are states of mind induced by "certain bodily actions" that are expressions of emotion, what caused the emotion in the first place? Not quite answering his own question, he pursues the thought that emotions are states of mind rather than actions.

Late marriages. Although the main theme of this passage is hereditary habits, he takes a moment for the "other principle," natural selection. If the custom of late marriage prevails, the human species will increase its longevity, since short-lived individuals will not reproduce. Darwin probably felt that he and Emma had made an adequate contribution to the general weal in this regard, since Charles was just turning thirty, and Emma thirty-one, when they married. Malthus' recommendation was for marriage between the ages of twenty-five and thirty as striking the right balance between the needs of society and those of the individual, the claims of restraint and the claims of passion.

> "Men are called 'creatures of reason,' more appropriately they would be *'creatures of habit.'* " [*N* 70]

pp. 68–73   *The imperfections of instinct.* Darwin has been making notes which constitute a running argument with an exponent of the view that the absolute perfection of animal instincts provides evidence for the existence of God via the argument from Design. Darwin, too, believed that animals are beautifully adapted to their environments and that instinctive behavior is one of the forms this adaptation takes. But his vision was of a changing world, and, change once granted, the fit of organism to world cannot remain absolutely perfect: "this not wholly true" best expresses the small but vital correction that Darwin makes throughout as he reads the Reverend Algernon Wells' lecture *On Animal Instinct*.

Wells argues that instinctive behavior is rigidly invariable, perfectly adapted to its end, directed to ends completely unknown to the behaving animal, and entirely divorced from human reason. "Not so!" cries Darwin—instinctive behavior is variable and imperfect. Animals have some sense of the goal toward which they are working—he is not quite sure of this and inserts a question, not to Wells but to himself: does a bird building a nest instinctively for *the first time* have any knowledge of the nest that will result from its actions?

For Wells' proposal that animals are "creatures of instinct" and men "creatures of reason" Darwin has *two* correctives. Animals are not entirely without reason, and men are guided less by reason than by habit.

Contrary to the alleged perfection of the relation between instinctive behavior and structure, Darwin had seen many anomalies during the voyage of the *Beagle*. Darwin paraphrases Wells: "Instincts & structures always go together: thus woodpecker." His answer: "but this is not so, the instincts may vary before the structure does; & hence we get over an apparent anomaly." (*N* 71)

For a fuller explanation of Darwin's meaning, we can turn to the *Origin of Species,* where he writes: "Can a more striking instance of adaptation be given than that of a woodpecker for climbing trees and for seizing insects in the chinks of the bark? Yet in North America there are woodpeckers which feed largely on fruit, and others with elongated wings which chase insects on the wing; and on the plains of La Plata, where not a tree grows, there is a woodpecker, which in every essential part of its organisation, even in its colouring, in the harsh tone of its voice, and undulatory flight, told me plainly of its close blood-relationship to our common species; yet it is a woodpecker which never climbs a tree!" (*Origin,* 184)

Darwin maintained his interest in this anomaly which only his theory could explain, and in 1870 while preparing the *Descent of Man* he finally got around to publishing a short "Note on the Habits of the Pampas Woodpecker (Colaptes campestris) ."*

---

* *Proceedings of the Zoological Society of London,* 1870, pp. 705–706. Reprinted in Barrett, *The Collected Papers of Charles Darwin, op. cit.*

In this expansion from the cryptic little remark in the N note-book, answering an obscure lecturer on the Argument from De-sign, to the authoritative, comprehensive statement in the *Origin*, we see the master hand of a scientific world traveler who kept his eyes peeled for the oddities, and the handwriting of a persistent collector of ideas and facts who kept his notebooks well.

> "instinct is hereditary knowledge of things which might be possibly acquired by habit. so bees in building cells, must have some means of measuring cells, which is a faculty." [N 77]

PP. 73–82

Faculties or instincts? This passage, vague and fragmented as it is, with quotations from an unidentifiable source, with pages missing, seems to represent an unnoticed turning point in Darwin's think-ing. In all the previous notes the word *faculty* has hardly oc-curred. The fundamental conception has been the analogy between instinct and habit, coupled with the proposition that habits can become instincts, thus providing a basis for evolutionary change.

Suddenly, we have a new choice to make—between instincts and faculties. Darwin is groping, and the distinction is not entirely clear. *Instinct* seems to refer to specific behavior patterns: bees making cells, birds migrating to a particular place. *Faculty* refers to something more like a general ability, useful in more than one situation: a bee's ability to "measure" the dimensions of the cells it is making, a watchmaker's ability or propensity to use his instru-ments, a bird's ability to find his way.

If one turns back just a few pages in this notebook, one dis-covers that the obscure Rev. Algernon Wells was, *inter alia,* making the distinction between instincts and faculties. But Darwin was mainly concerned to refute Wells's ideas about the invariability and perfection of instincts and did not take conscious note of the use-fulness of Wells's other point. Indeed, when he did notice it, his intent was to answer Wells, where the latter spoke of man as a creature of reason, for which Darwin wished to substitute habit.

Perhaps at the very end of his notes on Wells Darwin did begin to see the point, for in writing about the marvelous navigating abilities of various creatures, especially carrier pigeons, he writes: "this is class of so called instincts to which my theory no way ap-plies.—it is the acquirement of a new sense." (N 72) In Darwin's notes the phrase "my theory" varies in its referent; here he pre-sumably means that his theory of hereditary habits does not apply to faculties.

> "If we compare the judgments & actions of a young ani-
> mal with an old . . . we perceive great difference (& is
> not this difference same, but less in degree, as between
> man & child—) —what differs—Not instinct, for its charac-
> ter is invariability . . . some unnamed faculty—" [N 90]

pp. 87–
92

Not instinct? As we saw in the preceding paragraph, Darwin has
become wary of a psychology that leans too heavily on the concepts of habit and instinct. His dissatisfaction continues in this
groping, vacillating passage.

He considers momentarily the hypothesis that taste and preference might be accounted for by some sort of natural fit or harmony
between the organs of sense and the objects sensed. But he entertains the idea only to dismiss it in favor of the "great principle of
liking . . . *hereditary habit.*" (N 87) If specific ideas are to be
treated as inheritable, he may as well push the point as far as he
can: "A blind man might be born with the idea of scarlet." (N 88)

But he is not really satisfied with himself. Looking for a new
way of dealing with the problem of discovering the fundamental
terms of an adequate psychological theory, he proposes, "Get a
dictionary & make list of every word expressing a mental quality."
(N 88) Darwin was not the last psychologist to suggest that we
might clarify the properties of psychological reality by examining
the language used to describe it.*

Casting about for a psychological point of view, he reads Bell,
then Lamarck. He paraphrases Lamarck's idea that the role of
habit in behavior is inversely proportional to the role of intelligent
thought. One can see that this point will create trouble for Darwin. It means that "hereditary habit," his prime law of the evolution of mind, must itself evolve; it can do much work only when
the power of habit is strong. When that power gives way to increasing intelligence, the evolution of mind must take another path.
Darwin sees some such problem, for he says, lamely, that we cannot explain the faculty of reason as the outgrowth of habit and
then turn around and say that habits are the residues of reasoned
actions.

In spite of all these puzzles and doubts, Darwin still pursues the
hypothesis of hereditary habits. In one important sentence he
combines the two main ideas, natural selection and the inheritance
of habits: "could brain make a tune on the pianoforte, yes if every
individual played a little (& something destroyed bad brain)."

---

* See for example, Solomon E. Asch, "The Metaphor: A Psychological Inquiry," in *Documents of Gestalt Psychology,* edited by Mary Henle (Berkeley:
University of California Press, 1961).

(*N* 91) In other words, selection need not work only on accidental structural variations; it can also work upon variations produced by habits acquired during the lifetime of the individual.

The passage ends with still more puzzling, this time over the relation between ideas, actions, and emotions.

"George the lion is extraordinarily cowardly.—the other one nothing will frighten—hence variation in character in different animals of same species." [*N* 98]

PP. 93–100   *Variation and deviation.* These few pages are two sets of jottings, one probably written in June, the other in July 1839. There are some reading notes, from an unidentifiable source, on the physiology of emotions; there are some more observations Darwin made at the zoo of emotional expression in various animals. There is another reference to the idea of faculties, this time a description of the poet Cowper's power of imagination.

The fragmentary character of these notes is, of course, due to the fact that they are remnants—a number of pages have been excised. From what we know of Darwin's habits, it is likely that the missing pages had useful bibliographical references which he cut out and filed later on when he gave up keeping the little notebooks and switched to a filing system. In the final pages of the N notebook Darwin wrote on only one side of the page, possibly having in mind that this is more convenient for cutting and filing. These changes in his habits may have a meaning deeper than the mere mechanics of note-keeping. He may well have felt that he had mined this vein of ideas as far as he could profitably go. Except for a few miscellaneous notes on loose sheets, and the systematic notes on infant development he kept when his first child was born, these notes on the evolution of man and his mind just seem to peter out.

It would be nice to know just how he made the transition from his remark on variations in the personalities of lions to the topic of sexual deviation—but a half-page of notes is missing. The subject of deviation from a species norm was important to Darwin. When he first began his theoretical work, he treated variations as potential *evidence* for the occurrence of evolution. But after he developed the theory of evolution through natural selection, variation as a common event in nature was a necessary *assumption* in a causal system.

Just why should Darwin at this point have been struck by the notion that deviations of omission, such as he thought sexual continence to be, are less abhorrent than deviations of commission, such as suicide, abnormal sexual behavior, or—in some societies— cannibalism? What made him write the fascinating sentence in which he equates (as equally abhorrent to his fellow Englishmen) abnormal sexual actions and abnormal sexual impulses? These and other questions remain unresolved.

> ". . . man, a socialist, does not know other men by smell, but by looks, hence some obscure picture of other men, & hence idea of beauty. [N 109]

pp. 101–
end
What with excised and blank pages, long lapses between entries, and brief repetitions of thoughts made earlier at greater length, the remainder of the N notebook does not merit or permit much further discussion. There are a few points worth noting.

Insects and man. In the passage on pages 107–109, comparing social insects with other social animals, he makes three interesting points. First, in animals that are raised by their parents, the association between the pleasures of nurturance and the smell and look of the parents shapes the future social impulses of the young animal. Second, since man relies on vision and not on smell, the human baby forms "some obscure picture of other men, & hence idea of beauty"—in other words, our image of man begins to take shape in the cradle. Third, there is really a wide gulf between insects and vertebrates because the former do not rear their young. This last point echoes earlier remarks on the way in which complex behavior in insects is primarily instinctual, and in man primarily "intellectual." The point is repeated on page 115—the evolution of the bee's instinct and man's intellect are equally wonderful.

Instincts and learning. Typically, having for a moment made the distinction firm between instinct and reason, Darwin tries again to find the continuity between them. It will be helpful if instincts are amenable to some learned change. Darwin cites several examples—fear and flight reactions diminishing in birds and rabbits as they become accustomed to men and dogs. Darwin phrases this change as a case not of automatic habituation but of something like reason: "the birds at Maer have learned that he [man] is not dangerous." (N 112)

Questions. The notebook ends with many questions, and some plans for further study. He must read Hartley and Brown on psychology, Goldsmith, Hume, Adam Smith. Dogs dream—doesn't this mean that they daydream too? Are monkeys right-handed?

Does the flow of blood to the face increase during (sexual?) passion? "Do people of weak intellects easily fall into *habits*"?*

Pigeons. Darwin's experimental turn of mind is well exemplified by the following suggestion: "A carrier pidgeon carried & turned round & round in fainting state would it then know its directions"? In other words, if we prevent the pigeon from keeping track of its changes in position while being carried to the point of release, will it find its way home? Later on, when Darwin took up pigeon-fancying, his object was to come as close as he could to the experimental study of evolution, by breeding experiments. Along with his theoretical work, Darwin did an enormous amount of experimental work in many fields of science. In these notebooks we see that he had a number of ideas for psychological experimentation. Someday perhaps we will discover a notebook or a bundle of papers in which he wrote down the results.

---

* He probably meant mental retardates. In fact, the ability to form habits readily is greater in *more* intelligent people. The inverse relationship supposed by both Lamarck and Darwin is not borne out by modern research.

Old and Useless Notes

The transcriptions in this chapter are a collection of miscellaneous notes packaged originally in a single parcel and labeled by Darwin, "Old and useless Notes about the moral sense & some metaphysical points written about the year 1837 & earlier." Actually most of the notes, as indicated by their dates or their watermarks, were written during the years 1837–40. Among those which have no dates or watermarks are a few which Dr. Sydney Smith and Mr. Peter Gautrey of Cambridge University have dated from their knowledge of paper types, etc. The undated manuscripts seem also to have been written during this same period, as judged by their similarity in content to the others, and to the M and N notebooks.

"Old and USELESS *Notes about the moral sense*
& some metaphysical points
written about the year 1837 & earlier."

[OUN5][1] in Athenaeum
"Smart[2] Beginning of a New School Metaphysics,"—give my doctrines about origin of language—& effect of reason. reason could not have existed without it—quotes Ld Monboddo[3] language commenced in whole sentences.[4]—signs—?were signs originally musical!!! ??—

At least it appears all speculations of the origin of language.—
must presume it originates slowly—if their speculations are utterly
valueless—then argument fails—if they have, then language was
progressive.—

We cannot doubt that language is an altering element, we see
words invented—we see their origin in names of People.—sound of
words—argument of original formation—declension etc often show
traces of origin—

[OUN6][5] Mayo[6] Philosophy of living p. 264.
"Architecture is a fine amplification of two ideas in nature: a
developement of the thoughts expressed in Fingals cave, & in the
arched & leafy forest"
Very good!

[OUN7][7] I grant that the thrill, which runs through every fibre,
when one beholds the last rays of $\sigma\ \alpha$ or grand chorus are utterly inexplicable—I cannot <admit> think reason sufficient to give up
my theory—Viewing from eminence the wide expanse, of country,
netted with hedges & crowded with towns & thoroughfares, I grant
that man from the effects of hereditary knowledge, has produced

Spanish Fowl, Polish Fowl. "How does hen determine which most beautiful
cock?" Illustrations from *The Variation of Animals and Plants under Domestication.*

almost greater changes in the polity of nature than any other animal.

[OUN8][8] Aimé Martin[9] de l'Education des Mères Vol 1, p. 198.—
"Moralité, raison, beau ideal, infini conscience; voilà l'homme
separe de la matière et du temps! voila les facultes, q'il possede
seul sur la terre. J'ai trouvé son âme" etc—

Confesses these faculties of soul, treating of infinite not definable.
—His little chapter on each faculty of Soul.— (1) <Conscience>
moral sentiments imperative sense of duty—which makes struggle
in man.—two souls in one body[10]— (2) Beau ideal, refers chiefly to
moral, beau desires conscience & love.[11]—[With regard to ordinary
Beau ideal. Mem. Negro, beau,—Jeffrey denies all Beau.—How
does Hen determine which most beautiful cock, which best
singer—

Remember—avarice a compounded passion gained in life time][CD]
3. The Infinite.—*lives by hopes,* looks to eternity. (4) Reason,
some transcendental kind— (5) Conscience, not clear—Then these
last heads of separation between soul of man & intellect of beasts,
not clear.—?does not Mackintosh make great difference between
moral sense & conscience?—we admire what is right by one & are
ordered to do it by other.—

I suspect conscience an hereditary compound passion, like avarice.
Is there not something analogous to imperiousness of Conscience:
in Maternal instinct *domineering* over love of Master & sport etc
etc—The Bitch does not so act, because maternal instinct gives
most pleasure but because most imperious.—[12]

It would indeed be wonderful, if mind of animal was not closely
allied to that of men: when the *five senses* were the same[13]—In its
action—emotions—p. 176 & 177 good passage in French on what
dog dreams, awakes—does when Master takes Hat[14] de l'education
des Meres par L Aimé Martin

[OUN9][15] 1. Sensation is the ordering contraction (that is the only
evidence when consciousness is absent) in fibres united with nervous filaments.— plants? yes by distinct mechanism

2. Sensation of higher order where the sensation is conveyed over
whole body (which it may be in first case as when the excised
heart is pricked) and certain actions (only evidence where not
consciousness) are produced in consequence having some relation
to the primary sensation—man moving leg when asleep (or

habitual actions)—perhaps polypi—(so that lower animals are sleeping higher animals & not plants as supposed by Buffon) [16] Consciousness *is sensation No. 2.* with memory added to it, man in sleep not conscious, nor child—Evidence of consciousness, movements /?/ anterior to any direct sensation, in order to avoid it— beetles feigning death upon seeing object,—are Planariae conscious.—

Consciousness bears same relation to time & memory

[OUN10][17] Reynolds X discourse very curious as showing "the perfection of this science of *abstract* form" is the source of part of the highest enjoyment in mutilated statues[18]

[OUN11][19] Reynolds XIII Discourse (p. 115)
a very good passage about actions & decisions being the result of sagacity, or intuition, when individual cannot give reason, though he feels he is right—it is because each decision is made up of many partial results, & the impressions are then are <all> remembered, when the memory or reasons are forgotten. Our happiness etc, our well-being depend upon the "habitual reason,"—This power of the mind, faintly approaches to instinct[20]

[OUN11b][21] p. 142 "Upon the whole it seems"—"that the object of /all/ art is the realizing and embodying, what never existed but in the imagination."—[22]

Macculloch Vol. I. p. 115. Attributes of Deity—on Belief—you belief things you can give no proof for. & one often replies "what you say is perfectly true, but you do not convince me."—[23]
Belief allied to instinct—

In Elliotson's[24] Physiology much about sleep—How strange it, that Nature should have so little to do with art (p. 128) R. compares a view taken by camera obscura etc etc [illegible] How are my ideas of a general notion of everything applicable to the high idea /p. 131/ in [illegible] acting—My idea would make the mind have mysterious & *sublime* ideas independent of the senses & experience

p. 134 a painter must not an *action* or a scene in a garden.—yet both beautiful![25]

p. 136 Says Architecture does not come under imitative art.[26] (my view says yes. <old> mass of rock) or poetry my theory says yes imitating song—two primary sources, sight & hearing—

[OUN12][27] Staunton Embassy Vol II p. 405

Speculates on origin of sacrifices /common to many races/ thinks action toward /a king/ <man> <changed into> is carried on toward deity—& as king might like cruel pleasure, so sacrifices cruel.—

Something wrong here.—Origin is certainly curious.—

Chinese, S. American, Polynesians, Jews, Africans, all sacrifice. How completely men must have personified the deity.—

[OUN13][28] H. Tooke[29] has shown one chief object of language is promptness /of consequence/ newer languages become corrupt & whole classes of words /are abbreviations/ he thus derives from nouns & verbs—so that most EVERY language shows traces of anterior state??

[OUN14][30] Edinburgh Review Vol 18 (1st article)[31] on Taste /EXCELLENT/ Deficient in not explaining the possibility of handsome /ugly healthy/ young women, with good expression—statues not painted—music very good article—why flower beautiful ? can't children

[OUN15][32] S. Jenyns[33] Inquiry into the Origin of Evil
Review by Johnson in the Literary Magazine, 1756—Ceased in 1758. Read the Review or the Article.

[OUN16][34] A Planaria must be looked at as animal, with consciousness, it choosing food—crawling from light.—Yet we can split Planaria into three animals, & this consciousness becomes multiplied with the organic structure. it looks as if consciousness as effects of sufficient perfection of organization & if consciousness, individuality.—[35]

[OUN17][36] Quotes D. Stewarts System of Emotions.—T Mayo.[37] Pathology of the Human Mind Poor. on insanity.—Prevailing idea. owing to loss of *will*—chiefly excited by passive emotions.—Cannot quite perceive drift of Book.—Sympathy & affections chiefly fail.—Notices struggles <between> when insanity is coming on

(Thinks [distinct][38] analogy between dreams & insanity.)[39]

[OUN18][40] D Stewart[41] on the *Sublime*

The literal meaning of Sublimity is height, & with the idea of ascension we associate something extraordinary & of great power.—

2. From these & other reasons we apply to God the notion of living in lofty regions

3. Infinity, eternity, darkness, power, begin associated with God, these phenomena we (feel & ?) call sublime.—

4. From the association of power etc etc with height, we often apply the term sublime, where there is no real *sublimity*

5. The emotions of terror & wonder so often concomitant with sublime, adds not a little to the effect: as when we look at the vast ocean from any height.—

6. That the superiority & "inward glorying, which height by its accompanying & associated sensations so often gives, when excited by other means, as moral excellences, brings to our recollection the original cause of these feelings & thus we apply to them the metaphysical term sublime |

[OUN19] 7. So that in this Essay. D. Stewart does not attempt /by one common principle/ to explain the various causes of those sensations, which we call metaphorically sublime, but that it is through a complicated series of associations that we apply to such emotions this same term.—

Hence it appears, that when certain causes, as great height, eternity, etc. etc. produce an inward pride & glorying, (often however accompanied with terror & wonderment) which this emotion from the associations before mentioned, we call sublime.—

It appears to me, that we may often trace the source of this "inward glorying" to the greatness of the object itself or to the ideas excited & associated with it, as the idea of Deity with vastness of Eternity, which superiority we transfer to ourselves in the same manner as we are acted on by sympathy.

[OUN20][42] D. Stewart on taste

The object of this essay is to show how taste is gained how it originates, & by what means it becomes an almost instantaneous perception,[43]—Taste has been supposed by some to consist of "an exquisite susceptibility from receiving pleasures from beauties of nature & art"[44] But as we often see people who are susceptible of pleasures from these causes who are not men of taste & the reverse of this, taste evidently does not consist of this, but rather in the power of discriminating & respect good from bad.

And it is manifestly from this fact & the instantaneousness of the result, that the term taste is metaphorically applied to this mental power.[45] Although taste must necessarily be acquired by a long series of experiments & observations, & yet, like in vision, it becomes | [OUN21] so instantaneous, that we cannot ever perceive the various operations which the mind undergoes in gaining the result.

[OUN22][46] Lessing's Laocoon 2ᵈ Lect—The object of art, sculpture & painting, is beauty—which he thinks is a better definition than Winklemen's, who says it is simplicity with grandeur of character. —Hence Lessing shows expression of pain cannot be respected.[47] But what is beauty?—it is an ideal standard, by which real objects are judged: & how obtained—implanted in our bosoms—how comes it there? |

[OUN23] Laocoon p. 75
"The beauties developed in a work of art are not approved by the eye itself, but by the imagination through the medium of the eye;" he will allow the secondary pleasure of harmonious colours etc etc surely to be added.[48]

[OUN24] Lessings Laocoon
p. 125—says *new* subjects are not fit for painter or sculpture, but rather subjects which we know,[49] it is therefore the embodying of a floating idea,—as statue of beauty, is of the "beau ideal," my *instinctive* impression

[OUN25] September 6ᵗʰ 1838
Every action whatever is the effect of a motive.—[must be so, analyse[(a)] ones feelings when wagging one's finger—one feels it in passion, love—jealousy, as effect of bodily organisms—one knows it, when one wishes to do some action (as jump off a bridge to save

---

[(a)] one well feels how many actions are not determined by what is called free will, but by strong invariable passions—when these passions weak, opposed & complicated one calls them free will—the chance of mechanical phenomena.— (mem: M. Le Comte one of philosophy, & savage calling laws of nature chance)

another) & yet dare not, one *could* do it, but other motives prevent the action[50] see Abercrombie conclusive remarks p. 205 & 206.][CD] Motives are units in the universe.

[Effect of hereditary constitution,— education under the influence of others—varied capability of receiving impressions—*accidental* (so called like change) circumstances.[51]

As man hearing Bible for first time, & great effect being produced. —the wax was soft,—the condition of mind which leads to motion being inclined that way][CD]

one sees this law in man in somnabulism or insanity. free will (as generally used) is not then present, but he acts from motives, nearly as usual | [OUN26] difference is from imperfect condition of mind all motives do not come into play.—

†It may be urged how often one try to persuade person to change line of conduct, as being better & making him happier.—he agrees & yet does not.—because motive power not in proper state.—When the admonition succeeds who does not recognize an accidental spark falling on prepared materials.

From *contingencies* a man's character may change—because motive power changes with organization

The general delusion about free will obvious.—because man has power af action, & he can seldom analyse his motives (originally mostly INSTINCTIVE, & therefore now great effort of reason to discover them: this is important explanation) he thinks they have none.—

Effects.—One must view a wrecked man like a sickly one[p]—We cannot help loathing a diseased offensive object, so we view wickedness.—it would however be more proper to pity them to hate & be | [OUN27] disgusted with them. Yet it is right to punish criminals; but solely to *deter* others.—It is not more strange that there should be necessary wickedness than disease.

This view should teach one profound humility, one deserves no credit for anything. (yet one takes it for beauty & good temper), nor ought one to blame others.—

---

† A man may put himself in the way of contingencies.—but his desire to do arises from motives.—& his knowledge that it is good for him effect of education & mental capabilities.—

p Animals do attack the weak & sickly as we do the wicked.[52]—we ought to pity & assist & educate by putting contingencies in the way to aid motive power.—

   if incorrigably bad nothing will cure him

This view will not do harm, because no one can be really *fully* convinced of its truth, except man who has thought very much. & he will know his happiness lays in doing good & being perfect, & therefore will not be tempted, from knowing every thing he does is independent of himself to do harm.—

Believer in these views will pay great attention to Education— |

[OUN28] These views are directly opposed & inexplicable if we suppose that the sins of a man are, under his control, & that a future life is a reward of retribution.—it may be a consequence but nothing further.—

[OUN29] October 2ᵈ 1838
Those emotions which are strongest in man, are common to other animals & therefore to progenitor far back. (anger at the very beginning, & therefore most deeply impressed) shame perhaps an exception. (does it originate in a doubting feel between conscience & impulse) but shame /we alas know/ is far easier *conquered* than the deeper & worser feelings. Then bad feelings no doubt *originally* necessary revenge was justice.—No checks were necessary to the vice of intemperence, circumstances made the check—the licentiousness jealousy, & every one being married to keep up population. with the existences of so many positive checks.— (this is encroaching on views in second volume of Malthus) .[53] Adam Smith also talks of the necessity of these passions, but refers (I believe) to present day & not to wider state of Society.—Civilization is now altering these instinctive passions which being unnecessary we call vicious.— (jealousy in a dog no one calls vice) on same principle that Malthus has shown incontinence to be a vice & especially in the female

[OUN30] October 2ᵈ. 1838
Two classes of moralists: one says our rule of life is what *will* produce the greatest happiness.—The other says we have a moral sense.—But my view unites** both /& shows them to be almost identical/ & what *has* produced the greatest good /or rather what was necessary for good at all/ is the /instinctive/ moral sense: (& this alone explains why our moral sense <points> [prevents?] <is> to revenge). In judging of the rule of happiness we must

---

** Society could not go on except for the moral sense, any more than a hive of Bees without their instincts.—

look *far forward* /& to the *general* action/—certainly because it is
the result of what has *generally* been best for our good *far back*—
(much further than we can look forward: hence our rule may
sometimes be hard to tell)

The origin of the social instinct /in man & *animal*/ must be separately considered.—

The difference between civilized man & savage, is that the former
is endeavouring to change that part of the moral sense which experience (education is the experience of others) shows does not
tend to greatest good.—Therefore rule of happiness is to certain
degree right.—The change of our moral sense is strictly analogous
to change of instinct amongst animals.—[54]

[OUN31] Jan 13ᵗʰ 1839
My father received a letter from Mr Roberts /a person he had
known & directed many letters to/—could not <remember> read
Christian name; Fancied it looked like W. but concluded it could
not be so.—Looked at a direction book, but could not find out—
Directed his letter, & I observed he had written Wilson & pointed
it out; he was astonished, & said how very odd.—could not think
what had put Wilson into his head.—remembered, that he had
looked in direction book under head of Wilson, referred to Robert
& found his Christian name was *Wilson!!* How curious an inward,
unconscious memory.—[55]

[OUN32] Jan 14ᵗʰ 1839.—
My father says he has heard of many cases of ideots knowing
things, which are often repeated in a wonderful manner.—as the
hour of the day etc.—All habits must conduce to their health &
comforts.—Both ideots, old People & those of weak intellects.—

[OUN33] Westminster Review. March 1840[56]
p. 267—says the great division amongst metaphysicians—the school
of Locke, Bentham & Hartley, & the school of Kant & Coleridge is
regarding the sources of knowledge.—whether "anything can be
the object of our knowledge except our experience"—is this not
almost a question whether we have any instincts, or rather the
amount of our instincts—surely in animals according to usual
definition, there is much knowledge without experience. so there
may be in men—which the reviewer seems to doubt.[57]

1) [OUN34][58] Effects of Life in the abstract is matter united by certain laws different from those, that govern in the inorganic world; life itself being the *capability* of such matter obeying a certain & peculiar system of movements.[59] different from inorganic movements.—*

See Lamarck for this definition given in full.—[60]

According to the individual forms of living beings, matter is united in different modification, peculiarities of external form impressed, & different laws of movements.†

2) In the simplest forms of living beings namely /*one individual*/ vegetables, the vital laws act definitely (as chemical laws) as long as certain contingencies are present, (contingencies as heat light etc) *

During growth tissue unites matter into certain form;[62] invariable, as long as not modified by external accidents, & in such cases modifications bear fixed relation to such accidents.

But such tissue bears relation to whole, that is enough must be present to be able to exist as individual.— |

3) [OUN35] In animal growth of body precisely same as in plants, but as animals bear relation to less simple bodies, and to more extended space, such powers of relation required to be extended.

Hence a sensorium, which receives communication from without, & gives wondrous power of willing.[63] These †*willings* are common to every animal instinctive and unavoidable.

These willings have relation to external contingencies, as much as growth of tissue and are subject to accident; the sexual willing comes on period of year as much as inflorescence.—*

---

* Has any vegetable or animal *matter* been formed by the union of *simple* non-organic matter without action of vital laws.

† Hence there are two great <world> systems of laws /in the world/ the organic & inorganic—The inorganic are probably one principle for connect of electricity chemical attraction, heat & gravity is probable.—And the Organic laws probably have some unknown relation to them.[61]

* This is true as long as movement of sensitive plant can be shewn to be direct physical effect of touch & not irritability, which at least shows a local will, though perhaps not conscious sensation.

† (can the word *willing* be used without consciousness, for it is not evident, what animals have consciousness)

* I here omit the case (if such there are) of animals enjoying only movements such as sensitive plants. (But I include irritability for that requires will in part. ?Why more so than movement of sap or sunflower to sun? ∴ I should

4) The radicle of plants absorb by physical laws of endosmic & exosmic juices. arm of polypus, show either local or general will, & stomach likewise /does/.†

It is easy to conceive such movements & choice, & obedience to certain stimulants without conscience in the lower animals, as in stomach, intestines & heart of man.‡

How does consciousness commence; where other senses come into play, when relation is kept up with distant object, when many such objects are present, & when will directs other parts of body to do such.—

All this can take place & man not conscious as in sleep; or in sleep is man momentarily conscious, but is memory gone?—
Where pain & pleasure is felt where must be consciousness???*

5) [OUN36] Kirby thinks that there is one instinct to all animals modified according to species.⁶⁵ This I suppose he deduces from the ends in each case being the same, & the means very similar. It does not appear more than saying that the thinking principle is the same in all animals.

Kirby extends instincts to plants,⁶⁶ but surely instincts imply willing, therefore word misplaced.†

The meanings of words must be made out

Reason

Will

Consciousness

Definite instincts* begin acquired is most important argument, to show that they result from organization of brain; (:analogy:— as races are formed or modification of external form, so modifi-

---

think there were direct /physical/ effects of more or less turgid vessels; effect of heat, light or shade.)

   Joining two difficulties into one common one always satisfactory, though not adding to positive knowledge. lessening amount of ignorance |

† in Coralline are not two kinds of *life* vegetable and animal strictly united?

‡ How near in structure is the ganglionic system of lower animals & sympathetic of man.⁶⁴

* Can insects live with no more consciousness than our intestines have? |

† Eyton told me that his retriever Sailor he has seen push a hare through the bar of a gate before him, & then jump over the gate & bring it.

   Agrees with ONE animal

* not used by Kirby

cations of brain) As in animal no prejudices about souls, we have particular trains of thoughts as far as man; crows fear of gun.— pointers method of standing.—method of attacking peccari—re- triever—produced as soon as brain developed, and as I have said, no soul superadded,[67] so | 6) [OUN37] thought, however unintel- ligible it may be seems as much function of organ, as bile of liver. —? is the attraction[68] of carbon, hydrogen in certain definite pro- portions (different from what takes place out of bodies) really less wonderful than thoughts.[69]—One organic body likes one kind more than another—What is matter? The whole a mystery.—†

Instinct appear like hereditary memory; but first memory in many cases cannot be acquired by experience for child sucking.—And is it more wonderful that memory should be transmitted from gener- ation: than from hour to hour in <man> individual.‡

Acquired instincts analogous /& replace/ to experience gained by man in lifetime*

Hereditary memory not so wonderful as at first appears, & no too great advantage: for superiority of memory does not depend on its length: Many animals (as horses) very long & good memories— but on its multiplicity & the comparison of ideas.—

As man has so very few (in adult life) instincts.—this loss is com- pensated by vast power of memory, reason, etc, and many general instincts, as love of virtue, of association, parental affection—The very existence of mankind requires these instincts, though very weak so as to be overcome easily by reason.—Conscience is one of these instinctive feelings.†

6v) I think Pincher[71] shows surprise, walking home one day met him, with Mark[72] riding instantly followed, me and for five minutes every now and then howled.—Now I don't think this only pleasure: for it was different way of showing it, nor was there any cause, & if surprise was felt. analyse feelings.

---

† This materialism does not tend to Atheism.[70] inability of so high a mind without further end, just same argument. without indeed we are step towards some final end. production of higher animals—perhaps say attribute of such *higher* animals may be looking back, ∴ therefore consciousness, therefore re- ward in good life
‡ Perhaps even the most complicated instinct might be analysed into steps, as species change.—Must be so if Lamarck's theory true
* But habits acquired even by <children> plants!
† As sexual instinct comes on late in life, man almost alone in this case can perceive instinct. boy takes delight in mammae before any reason had told him this distinctive mark. it is downright instinct, leading to touch a particular organ.— |

Mr. Wynne[73] says, that beyond doubt *courage* is hereditary in fowls & not effect of feeling of individual force in any individual. —His Malay breed /of fowl/ totally different habits from Europaean. begin to prowl about in the evening /seldom leave their perch till evening/ crow different.—*Hereditary* effect of former tropical climate ((analogous to inflorescence of Tropical plants when imported & plants sleeping)) ((good show aquirement or obliteration of instincts)) |

7) [OUN38] As definite instincts modified by hereditary succession so perhaps general ones.—Parental feelings weakened in Otahiti: fear of death in Hindoo population.—Slightly modified in many countries, hence national character, love of country, of association etc stronger in some than others—Hence superiority of Christian over Heathen race.—But as no great modification in brain would probably take place without corresponding change in /external/ man; and as all men nearly same species, so general instincts nearly same; which argument probably applies to particular instincts of animals, even in wild state; certainly to the domesticated.—†

General—Instincts certainly appear a sort of acquired memory. a permanent secretion of thought. (or under contingencies of stimulants of certain kinds such secretion) *

This memory† especially the general kind taking pleasure in virtue because acquired in past ages, seems to indicate that when we turn into angels, this imperfect memory may become perfect & we may look back to definite action or to our conscious selves.— ((Such memory may go back to animals which were changed into man ∴. they meet their reward!))

The difference between hereditary memory & individual secretion of thought, may be no more /different/ than sexual intercourse in plants is involuntary, in man voluntary: ?False.—secretion in both involuntary, <application> ejection only his will: there must be case of secretion being some time governed by will in some animals, involuntary in others. |

---

† NB. Two dogs having very different instinct always obtain peculiarities of external configuration
* or an ASSOCIATION of pleasures with certain actions performed by your parents, conscience
† Perhaps should hardly be called memory; you cannot call the frame of mind which makes music pleasant, a memory; yet that frame is enhanced by memory of what has been heard; so love of virtue enhanced by this hereditary kind of memory.—

1) [OUN39][74] Why may it not be said thought perceptions will, consciousness, memory, etc. have the same relation to a living body (especially the cerebral portion of it) that attraction has to ordinary matter.

The relation of attraction to ordinary matter is that which an action bears to the agent. Matter is by a metaphor said to attract; & hence if thought, etc bore the same relation to the brain that attraction does to matter, it might with equal propriety be said that the /living/ brain perceived, thought, remembered, etc*

Now this would certainly be a startling expression, & so foreign to the use of ordinary language that the onus probandi might fairly be laid with those who would support the propriety of the expression. They would do well to ask themselves the converse ((because there are living bodies without these faculties)) of the question above stated, & indeed until we know what answer they would give in support of their view it is impossible to shew satisfactorily its erroneousness. ((it is point of indifference)) |

2) In the absence of such a guide we can only point out the mode /of perceptive action/ by which we come to conceive of matter as attracting & shew that the groundwork <of this> is entirely wanting by which thought or memory might be in like manner attributed to brain.

There are two modes of perceptive action by which bodily action is made known to us, revealing respectively what are called its subjective & objective aspect.
The subjective aspect of bodily action is revealed to us by the effort it costs us to exert force or by internal consciousness; the objective, by our *external* /what/ senses in the way in which we apprehend the *force* of inanimate bodies. How we identify the two aspects as different phases of the same object of thought is a question which ought to be clearly comprehended by anyone who wishes to fully understand this subject, but the answer to it would require a considerable degree of attention. ((How do the senses affect us except by internal consciousness—)) |

3) [OUN40] We must endeavour to do without it as well as we can <The objective aspect of Bodily> action ((as recognized by our external senses)) consists in the *manifestation* of force /i.e. movement?/ capable of being traced to the body of the individual to whom the action is attributed: force (be it remembered) being a

---

* Well the heart is said to feel

phenomenon apprehended by the same faculty with matter & being necessarily exhibited in & by matter.

The phenomena of gravity considered in themselves consist in a force manifested in every particle of matter directed towards every other particle; but FORCE, <objectively> considered, is a phenomenon the essence of whose existence consists in its communication to other matter in the course of its DIRECTION, & thus when we apprehend force in inanimate matter we feel dissatisfied until we can point out | 4) the source from which it arises. ((How can force be recognized by our external senses—only movement can))

But coming round to the <subjective> aspect of action ((as known by the exertion of our own power & consciousness of it)) we are conscious that we ourselves can originate in any point an opposition of forces balancing each other & moving in opposite directions. We are satisfied therefore if we can trace any force in inanimate matter up to the action of some animated agent Now the phenomena of gravity are manifestly the same as if every particle of matter were an animated being pulling every other particle by invisible strings & as on this supposition the forces manifested would be fundamentally accounted for, we prefer this metaphorical mode of stating the fact to the mere statement of the <force exhibited in every> phenomena actually apprehensible by sense. |

5) [OUN41] There is nothing analogous to this in the relation of thought, perception, memory etc. either to our bodily frame or the cerebral portion of it

Thoughts, perceptions etc., are modes of subjective action—they are known only by internal consciousness & have no objective aspect. If thought bore the same relation to the brain that force does to the bodily frame, they could be perceived by the faculty by which the brain is perceived but they are known by courses of action quite independent of each other. A person might be quite familiar with thought & yet be ignorant of the existence of the brain. We cannot perceive the thought of another person at all, we can only infer it from his (its) behaviour. ((attraction of sulphuric acid for metal))

Thought is only known *subjectively* ((??)) the brain only objectively ((We do not know attraction objectively)) |

6) The reason why thought etc. should imply the existence of something in addition to matter is because our knowledge of mat-

ter is quite insufficient to account for the phenomena of thought. The objects of thought have no reference to place.
[We see a particle move one to another, & (or conceive it) & that is all we know of attraction. but we cannot see an atom think: they are as incongruous as *blue* & *weight:* all that can be said that thought & organization run in a parallel series, if blueness & weight always went together, & as a thing grew blue it /uniquely/ grew heavier yet it could not be said that the blueness caused the weight, anymore than weight the blueness, still less between [illegible] [them?] so different as action thought & organization: But if the weight never came until the blueness had a certain intensity (& the experiment was varied) then might it now be said, that blueness caused weight, because both due to some common cause: —The argument reduced itself to what is cause & effect: it merely is /invariable/ priority of one to other: no not only thus, for if day was first, we should not think night an effect]^{CD} ((Cause and effect has relation to forces & mentally because effort is felt))

1) [OUN42] May 5th, 1839. Maer
Mackintosh Ethical Philosophy
On the Moral Sense
Looking at Man, as a Naturalist would at any other Mammiferous animal, it may be concluded that he has parental, conjugal and social instincts, and perhaps others.*—The history of every race of man shows this, if we judge him by his habits, as another animal. These instincts consist of a feeling of love <& sympathy> or benevolence to the object in question. Without regarding their origin, we see in other animals they consist in such active sympathy that the individual forgets itself, & aids & defends & acts for others at its own expense.—Moreover any action in accordance to an instinct gives great pleasure, & such actions being prevented by <necessarily> some force give pain: for instance either protecting sheep or hurting them.—Therefore in man we should expect[76] that acts of benevolence towards fellow <feeling> creatures, or of kindness to wife | 2) [OUN43] and children would give him pleasure, without any regard to his own interest. Likewise if such actions were prevented by force he would feel pain. [By a very slight change in association if others injured these objects, without his being able to prevent it, he would likewise feel pain.—If he saw another man acting in accordance to his instincts, <he would know the many experiences pleasure> & by association he would feel part of the pleasure, which the actor received.—If either man

* p. 113 Mackintosh[75] Grotius has argued nearly so

did not obey his instincts from interference of passion, he would
feel pain, which would generally be anger, as he would be tempted
to interfere, but with respect to himself it would be remorse as will
be presently shown.—This then is moral approbation, as far as it
goes.]CD77 But should he [be] prevented by some passion or appetite, what would be the result? In a dog we see a struggle between
its appetite, or love of exercise & its love of its puppies: the latter
generally soon conquers, & the dog | 3) [OUN44] probably thinks
no more of it.—Not so man, from his memory & mental capacity
of calling up past sensations he will be forced to reflect on his
choice: an appetite gratified gives only short pleasure. passion in
its nature is only temporary, & we do not afterwards think of it.[78]
—Whatever the cause of this may be, everyone must know, how
soon the pleasure from good dinner, or from a blow struck in passion fades away, so that when man afterwards thinks why was
such an instinct not followed for a pleasure now though so trifling
he feels remorse.—*

He reasons on it & determines to act more wisely other time, for he
knows that the instinct (or conscience) is always present (which
is indeed, often felt at very time it is disobeyed) & is sure guide.—
Hence conscience is improved by attending & reasoning on its action, & on the results following our conduct.—If the temptation to
disobey the conscience is extremely great | 4) [OUN45] as starva-
tion, or fear of death, one make allowance & either excuses the
/non-/ following of one's conscience, & palliates the offence; one
always admires the habit formed by <conscience> /obedience to
instinct/, or rather strengthened instinct, even when our reason
tells[79] us the action was superfluous, as one man trying to save another in desperation.—This shows, that our jealousy, that the instinct *ought* to be followed, is a consequence of that being part of
our nature, & its effects lasting, whilst passions although equally
natural leave effects not lasting. By association one gains the rule,
that the passions & appetite should /almost/ always be sacrificed
to the instincts.—One does not feel it is wrong in very young child
to be in a passion, any more than in an animal—which shows that
it is owing to some <subsequent> power[80] (reason) obtained by
age, which should show the child, which of its instincts are best to
be followed.—Yet even at this time, malevolence, when not urged
to it by passion, shows a bad child.—

Hence there are certain instincts pointing out lines of conduct to
other men, | 5) [OUN46] which are natural (& which /when

---

* The cause perhaps lies in its frequency & in its consisting in desire gratified
& therefore as soon as desire is fulfilled, pleasure forgotten

present/ give pleasure) & which man *ought* to follow—it is his duty to do so.—So we say a pointer *ought* to stand—a <spaniels> housedog's duty is to watch the house.—it is part of <duty> their nature.—When a pointer springs his bird, one says for shame (& the /old/ dog really feels ashamed?) not so puppy, we <do> try to teach him & strengthen his instincts.—so man *ought* to follow[81] certain lines of conduct, <although> even when tempted not to do so, by other natural appetites he is *monster,* or unnatural if malevolent, or hates his children without some passion.—If his passions strong & his instincts weak, he will have many struggles, & experience only will teach him, that the instinctive feeling in its nature being always present, & his passions shortlived, it is to his interest to follow the former; & likewise then receive the normal approbation of his fellow men.— | 6) [OUN47] Hence man* must have a feeling, that he *ought* to follow certain lines of conduct, & he must soon *necessarily* learn that it is his interest to follow it, even when opposed by some natural passion.—[a]

By interest I do not mean any calculated pleasure but the satisfaction of the mind, which is /much/ formed by past recollections.[83] —Hence he has the right & wrong in his mind.—Now we know it is easy by association to give /almost/ any taste to a young person, or it is accidentally acquired from some trifling circumstance.— Thus a child may be taught to think almost anything nasty (<accidentally> /by old association/ comes to this conclusion not owing to peculiarity of organ of taste, for when grown up often conquers it).[84] It will be only rarely that it thinks that nasty, which the *natural* tastes say is good. Yet horseflesh shows that even this is possible.—So that there [is] nice & nasty in taste, & right & wrong in action, so a child may be taught, or will acquire from seeing conduct of others, the feeling that *almost* (rarely if opposed to *natural* instincts) any action is either right or wrong.— |

7) [OUN48] Hence what parents think will be good for the child on the long run, & for themselves & others (as the parents are instinctly benevolent) they will teach to be wrong or right; this teaching may be curiously modified by circumstances of country, so will the conscience in these cases.—Those instructions, which the child sees uniformly performed by the teachers & all around

---

* The conscience rebukes malevolent feelings, as much as actions, therefore Sir J. M. talks too much about the contiguity to will.[82]
(a) The origin of passions too strong for our present interest receive simple explanation from origin of man.—

him, will be paramount, hence the law of honour.* & the etiquettes of Society—Anyone who will reflect must feel, how like to injured conscience, is the feeling of any custom of society broken.—& how far more acute the feeling really is.—all these associated /habitual/ feelings become like instinctive ones, <which either lead to actions or not, as feeling of cowardice /this is not connected with sense/> instantaneous declaring it is right or wrong.—/[just as in taste of the mouth]^{CD}/†

Feelings of the mind, whether leading to action or not, are the parts of our nature, subject to their instincts, & associations.— often feelings which do not lead to action are repressed Thus avarice etc etc.— |

8) [OUN49] In the beginning I mentioned only three instincts,—* I am far from saying there are not more, or that the three are as simple as I have said.—the social instinct may be combined with feeling towards one as a leader.—the conjugal feeling may be directed towards one or more.—It will be hard to discover this, for the different races of man may have different instincts, as we see in dogs and pidgeons.—But as man is animal at head of series in which /special/ instincts decrease, I should think they were very few & general in their nature.—So that we have some it is sufficient to give rise to the feeling of right & wrong.—on which /almost/ any other might be grafted.—

Origin of the instincts
Hartley, (according to Sir J.) /p. 254, etc etc/ explains our love of another, as pleasure arising from association from having received benefits from this person.—But the love is instinctive, & how does it apply to mother loving child, from whom, she has never received any benefit.—Yet I think there is much truth in doctrine, for* |
9) [OUN50] we can thus explain love of place.—although here we have not received pleasure *from* the place, but merely *in* the place.[86] & yet place calls up pleasure.—This feeling seems to vary in races of man, & certainly in /species of/ animals in which case

---

* Sir J. M. gives different explanation of law of honour from Paley
† My theory of instincts, or hereditary habits fully explains the cementation of habits into instincts.
* Instinctive fear of death: of hoarding . . Ld Kames which Sir J. says is so ridiculous[85]
* the instinct of sociability & sociability, doubtless grow together

it undoubtedly is instinctive. But does not Hartley explanation apply perfectly to origin of these instincts.—the having received pleasure from some *one* /person/ in early infancy, during many generations giving love of mother: the having received some advantage from man during many generations giving the social feelings.—*

Although I cannot pretend to say how far & minutely our instincts extend, †yet as they are acquired by social animals, living under certain conditions, in this world, they <will conform to the law> can only be such, as are consistent with social animals, that is which have a beneficial tendency, (not to any one individual, but to the whole past race) <I cannot> <no one> doubts) | 10) [OUN51] that the instincts of bees & beavers /deer/ have <been formed> a beneficial tendency to them as <social> animals of peculiar <kinds> social feelings, & living under certain conditions; by my theory they have been formed by the circumstances, which have led to the peculiarities, & hence <must> only that <have> which had a beneficial tendency during past races could become instinctive.—‡

Better simply put it, beneficial tendency in every instinct to the species in which it occurs [or, more correctly /in which it/ has been so in some past time, hence passions]CD /although perhaps useful at present to some extent./ Hence this is the law of our instinctive feelings of right & wrong,—education of parents strives*

---

* According to my theory, all instincts demand some explanation

† *On Law of Utility* Nothing but that which has beneficial tendency through many ages could be acquired, & we are certain from our reason, that all which (as we must admit) has been acquired, does possess the beneficial tendency)

‡ It is probable that becomes instinctive which is repeated under many generations (& under unknown conditions) (for pig will not so readily attain *instinct* of pointing as a dog.—also, age has much influence.) —& only that which is beneficial to race, will have reoccurred. NB. Until it can be shewn, what things easiest become instinctive, this part of argument fails, or rather is weak.—

* for it strives to give conduct beneficial to all the children,† (each himself) & parents, & hence to nearly all the world—

† Our tastes in mouth by my theory are due to <habit> hereditary habit (& modified & associated during lifetime) , as in our moral taste. [NB. footnote of a footnote]

p. 152 Reason can never lead to action.—87

p. 164 Ld Shatsbury [i.e., Shaftsbury] under term of *Reflex Sense* seems to have <compared> perceived the comparison between our instinctive feelings & our short lived Passions.

As emotions change, from civilization, education changes, & probably likewise instincts, for the same law effects both.—

changes /in accordance to beneficial tendency/ will most readily effect the instincts, for they are in accordance with it. thus a dog may be trained to hunt one pig sooner than other, rather than change hunting instinct.

to same end.—& general actions of community must frequently teach same end.—Hence this becomes the law of right & wrong, though that part which is acquired by association from education & imitation, has often been perverted from want of reason.—Hence as Eugenius[88] says, slow growth of rule of right.—

10v) State broadly in child or animal it is equally proper to obey anger as benevolence (but not cool malevolence). it only after reason comes into play that anger can be said to be wrong.—for then only is it perceived, that our passions are too strong for our instincts to gain long-lived good, ie happiness, yet this system not *selfish.—explained by principles of Mackintosh.—*

p. 262 Some good remarks on analogy of pleasure of imagination /the utility part being blended & lost/ & moral sense—my theory explains both, perhaps, by habit—[89]

1) [OUN52] Whewells preface
It appears that Sir J. & others think there is distinct faculty of conscience.—I believe that certain feelings & actions are implanted in us, & that doing them gives pleasure & being prevented uneasiness, & that this is the feeling of right & wrong.—so far it has *independent* existence. & is supreme because it is a part of our nature which regulates our feelings steadily & not like an appetite & passion, which receive enjoyment from gratification & hence are forgotten—only so far do I admit its *supremacy**

p. 37 Whewell gives Mackintosh's theory: the remarks about "contact with will" is unintelligible to me.—conscience regulates feelings, as of cowardice.—the whole appears to me rather rigmarole. —He does not say anything about any principles *born* in us.[91]— Great difference with my theory.—see p. 349—remark on this point.—*

11v) p. 224 Hume's Inquiry—good abstract of Butler & arguments of beneficial tendency of affections.—If ever I write on these subjects consult <following> pages. <p. 231> marked in my Mackintosh |

---

* p. 194 etc. Butler's view given on conscience; I cannot admit it.—see notes by me.[90]

* p. 333 & 377 some remarks showing that instinct cannot be said to guide *will*, as bird building nest, but supplies it—instinctive feelings will doubtless lead to similar actions which in prior generations led to their formation.— N.B. feeling or emotion rises from hereditary action on body,—this feeling, when instinctive will lead to action.—the passion rising from weariness leads to striking blows.—[92]

1) [OUN53] *Mackintosh's Ethical Philosophy*

p. 6—"The pleasure which results when the object is attained (the gratification of one's offspring) is not the aim of the agent, for it does not enter into his contemplation"—Now Eugenius[93] would contend against this—but the pleasure a dog has in obeying its instinct, as young pointer to point—clearly shows this is true.

p. 13—Affections cannot be analysed with "power" etc etc etc—& if termed "selfish," must be subclassed as "disinterested"[94]

p. 14 It is *allowed,* that we have conception of moral obligation /when grown up ? ? ?/ & the question is, whether this can be resolved into some operation of intellectual faculties[95]—Will Eugenius[96] allow this moral obligation? |

2) [The improvement of the instinct of a sheperd dog, is strictly analogous to education of child,—causing many actions to be considered right & wrong,—to be associated with the approving or disapproving instinct— which were not originally, if the shepherd dog had no instinct to commence with scarcely possible to teach it—all *dogs* might be taught, but not *cat,* that is not act by gusto, though by fear it might be partly made.]^CD

p. 21 "Why ought I to keep my word"[97]—gives the problem of ethics—[my answer would be to all such cases—either, that from the necessities /& good/ of society such conduct is instinctive in me (& as a consequence, but not cause gives me | 3) [OUN54] pleasure) or that I have been taught or habituated to associatical, the emotions of this instinct, with that line of conduct, & if taught rightly, it will be for the general good, that is the same cause which gives the instinct.—]^CD

p. 22 says affections, desires, & moral sense all different.—

p. 22 Butler & Mackintosh characterize the moral sense by its "supremacy."[98]—I made its supremacy, solely due to greater duration of impression of social instincts, than other passions, or instincts.—is this good?—I should think some parts of the emotive part of man, may be quite artificial, as avarice love of *gold.*— love of fame—Yes Hartley explains this & Mackintosh shows the change produced— |

4) p. 38 Conscience checks the *wish* to outward gratification, whilst no desire of gratification will check the consciences desire for virtue.—[I expect there is some fallacy here.—at least point of /false/ honour will stop all wish to gratify <it> anything contrary to it]^CD NB. the very end of conscience is stop to [i.e., to stop]

wishes of passion etc. Whilst the passions have no relation I think this is nonsense—My theory of durableness will explain it.—

Would not the maternal affections (in a dog & therefore not <instinct> conscience) equally destroy all wish of outward gratification.—see what cases Mackintosh gives & try it.— |

5) [OUN55] p. 241 (1) Any action by habit may be thought wrong.—& conscience will imperiously say so, & produce shame & remorse[99]—[Thus pungency of one's feeling for indecency—preposterously so, for Marquesans think only of prepuce, crepitando,]CD & *if passion makes one break these artificial rules, get remorse*—((hence desires do not intervene between this kind of conscience & the will, though this conscience does between the desires & will?)) (2) It is other question what it is desirable to be taught,—all are agreed general utility (3) It is other question whether any thing is taught instinctively; I say yes, & my explanation agrees with last head.— (4) It is other question, how the feeling of ought, shame, right & wrong comes into mind in first case—seeing how shame is accompanied by blushing, bears some relation to others |

if so, it is perhaps deviation from the *instinctive,* right & wrong.— (animals excepting domesticated ones have no right & wrong except instinctive ones) . Perhaps my theory of greater permanence of social instincts explains the feeling of right & wrong—arrived at first <rationally> by feeling—reasoned on, steps forgotten, habit formed,—& such habits carried on to other feelings, such as temperance, acquired by education.—In similar manner our *desires* become fixed to ambition, money, books etc. etc. <]> the "secondary passion" of Hutcheson unfolded by D. Hartley.[100]

Notes by Paul H. Barrett

1. 1838 watermark.

2. Smart, B. H., *Beginnings of a New School of Metaphysics: Three Essays in One Volume: Outline of Sematology.—MDCCCXXXI. Sequel to Sematology.—MDCCCXXXVII. An Appendix, Now First Published.* Richardson, London, 1839, pp. 3–5.

3. Monboddo, James Burnett (1714–1799) , Scottish judge, philosopher, and author of *Of the Origin and Progress of Language,* 6 vols., Kincaid and Creech, Edinburgh, 1773–1792.

4. Smart, *op. cit.,* n., pp. 26–27.

5. 1838 watermark.

6. Mayo, 1838, *op. cit.,* p. 264: "Nature is beyond art. For Nature is divine art. Yet human art may select and combine her elements, and reproduce some of her conceptions. Architecture is a fine amplification of two ideas in nature: a develope-

ment of the thoughts expressed in Fingal's cave, and in the arched and leafy forest. To learn its powerful influence on the imagination, let any one visit York Cathedral, for an interior;—or, which is not less deeply moving, view in bright moonlight, at some silent hour, the magnificent elevation of St. Paul's."

7. Watermark, c. 1838. In the passage σ and α represent wavy lines drawn by Darwin.

8. Date not traced.

9. Aimé-Martin, Louis, *De l'éducation des mères de famille; ou, de la civilisation du genre humain par les femmes*, 2 vols., Meline, Brusselles, 1837, p. 198.

10. Aimé-Martin, Louis, *The Education of Mothers: or the Civilization of Mankind by Women*, transl. from the French by Edwin Lee. Revised from the Fourth French Edition, Whittaker & Co., London, 1842: "There exists, then, two wills in man: there is only one will in animals. Man therefore is alone free upon the earth. He alone can struggle with and conquer himself. He alone escapes from the fatalities of organization."

11. *Ibid.*: "The type of the beautiful is immutable—eternal; it exists; for we have the consciousness and the love of it: consciousness, to incline us to seek it; love, to render us worthy of contemplating it."

12. *Ibid.*: ". . . the moral sense is not dependent upon our intelligence, and . . . it imperatively indicated to us what we must do in order to deserve happiness."

13. *Ibid.*: "Observe this dog, asleep at my feet; the nerves of his brain are distributed to the organs of the five senses. . . ."

14. *Ibid.* "Here is my dog, asleep in the chimney-corner; his sleep is disturbed, he is dreaming of pursuing his prey, he attacks his enemy, he sees him, he hears him; he has sensations, passions, ideas. When I rouse him up, his visions disappear, and he becomes calm; when I take up my hat he darts out, jumps about, looks me in the face, and studies my actions; he crouches at my feet, runs to the door, is joyful or sorrowful according to the will which I express. . . . Here is an animal who thinks, wills, remembers, and combines his ideas. There are moments in which I am tempted to believe him to possess a soul, for, in fact, I find in his intelligence the phenomena which exist in my own. . . ."

15. 1837 watermark.

16. *Natural History, General and Particular, by the Count de Buffon*, transl. by William Smellie, 3rd ed., 9 vols., Stahan and Cadell, London, 1791, vol. 2, p. 7: "If the sensation of an oyster, for example, differ in degree only from that of a dog, why do we not ascribe the same sensation to vegetables, though in a degree still inferior? This distinction, therefore, between the animal and vegetable, is neither sufficiently general nor decided."

P. 8: "[Thus] there is no absolute and essential distinction between the animal and vegetable kingdoms; but . . . nature proceeds by imperceptible degrees from the most perfect to the most imperfect animal, and from that to the vegetable: Hence the fresh water polypus may be regarded as the last of animals, and the first of plants."

17. 1837 watermark.

18. Reynolds, Joshua, *The Literary Works of Sir Joshua Reynolds, to which is Prefixed a Memoir of the Author by H. W. Beechy*, 2 vols., Cadell, London, 1835, Vol. 2, pp. 8–9: ". . . what artist ever looked at the Torso without feeling a warmth of enthusiasm, as from the highest efforts of poetry? From whence does this proceed? What is there in this fragment that produces this effect, but the perfection of this science of abstract form?

"A mind elevated to the contemplation of excellence, perceives in this defaced and shattered fragment, *disjecta membra poetae*, the traces of superlative genius, the reliques of a work on which succeeding ages can only gaze with inadequate admiration."

19. Late 1838.

20. Reynolds, 1835, *op. cit.*, Vol. 2, p. 62: ". . . our conduct in life, as well as in the Arts, is, or ought to be, generally governed by this habitual reason: it is our happiness that we are enabled to draw on such funds. If we were obliged to enter into a theoretical deliberation on every occasion, before we act, life would be at a stand, and Art would be impracticable."

21. Late 1838.

22. Reynolds, 1835, *op. cit.*, Vol. 2, p. 78: "Upon the whole, it seems to me, that the object and intention of all the Arts is to supply the natural imperfection of things, and often to gratify the mind by realising and embodying what never existed but in the imagination."

23. Macculloch, *op. cit.*, Vol. 1, p. 115.

24. Elliotson, John, *Human Physiology . . . With which is incorporated, much of the elementary part of Institiones physiologicae of J. F. Blumenbach . . .* 5th ed., Longman, etc., London, 1840 [1835–40] [section on sleep, chapter 27, pp. 598–698].

25. Reynolds, 1835, *op. cit.*, Vol. 2, p. 73: ". . . no Art can be grafted with success on another art . . .

"If a Painter should endeavor to copy the theatrical pomp and parade of dress, and attitude, instead of that simplicity, which is not a greater beauty in life than it is in Painting, we should condemn such Pictures, as painted in the meanest style.

"So, also, Gardening, as far as Gardening is an Art . . . is a deviation from nature . . ."

26. *Ibid.*, p. 74: "Architecture . . . applies itself, like Music (and, I believe, we may add Poetry) , directly to the imagination, without the intervention of any kind of imitation."

27. October, 1838.

28. The watermark on this scrap of paper is, "[What] man—33."

29. Tooke, John Horne, *Epea Pteroenta, or the Diversions of Purley. Revised and corrected with additional notes, by Richard Taylor*, Tegg, London, 1860, p. 14: "The first aim of Language was to communicate our thoughts; the second to do it with dispatch. Pp. 23–24, "In English, and in all Languages, there are only *two* sorts of words which are *necessary* for the communication of our thoughts. . . . 1. Noun, and 2. Verb."

30. Date not traced.

31. Alison, Archibald, "Essays on the Nature and Principles of Taste," 2 vols., Edinburgh, 1811, 830 pp. [Review], *The Edinburgh Review, or Critical Journal:* for May 1811. . . . Aug. 1811. Vol. 18:1–46, 1811, p. 10: "The most beautiful object in nature, perhaps, is the countenance of a young and beautiful woman . . . what we admire is not a combination of forms and colours . . . but a collection of signs and tokens of those feelings and affections, which are universally recognized as the proper objects of love and sympathy."

P. 17: "The forms and colours that are peculiar to [children], are not necessarily or absolutely beautiful in themselves; for in a grown person, the same forms and colours would be either ludicrous or disgusting."

P. 18: "Take, again, for example, the instance of female beauty,—and think what different and inconsistent standards would be fixed for it in the different regions of the world;—in Africa, in Asia, and in Europe;—in Tartary and in Greece;—in Lapland, Patagonia and Circassia. If there was anything absolutely or intrinsically beautiful, in any of the forms thus distinguished, it is inconceivable that men should differ so outrageously in their conceptions of it: If beauty were a real and independent quality, it seems impossible that it should be distinctly and clearly felt by one set of persons, where another set, altogether as sensitive, could see nothing but its opposite . . ."

Pp. 18–19: "The style of dress and architecture in every nation, if not adopted from mere want of skill, or penury of materials, always appears beautiful to the natives, and somewhat monstrous and absurd to foreigners . . . the fact is still more striking, perhaps, in the case of Music . . ."

32. Date not traced.

33. Jenyns, Soame, *A Free Inquiry into the Nature and Origin of Evil*. Dodsley, London, 1757.

34. 1837 Watermark.

35. In the margin beside this passage Darwin drew a vertical pencil mark, and two large question marks.

36. 1838 Watermark.

37. Mayo, Thomas, *Elements of the Pathology of the Human Mind*, Murray, London, 1838.

38. This word is difficult to decipher.

39. Mayo, Thomas, *Elements of the Pathology of the Human Mind*, Murray, London, 1838: ". . . much of my reasoning . . . flows similarly to that by which the operations of sleep and dreaming are explained by Mr. Dugald Stewart. These states have, in truth, always appeared to me to possess a striking affinity to that of the insane,—with this important difference, that there is a constant readiness in the mind to be roused out of sleeping or dreaming state, but no such readiness to emerge out of insanity, when once incurred. And again, that in sleep every voluntary action is suspended; whereas in madness, the will acts with considerable force, though in a more limited extent, than in the sane state."

40. Date not traced.

41. Stewart, Dugald, *The Works of Dugald Stewart*, 7 vols., Hilliard and Brown, Cambridge, 1829, Vol. 4, "Essay Second, On the Sublime," pp. 265–317. Darwin outlines here the main principles discussed by Stewart in this Essay.

42. Date not traced.

43. Stewart, *op. cit.*, Vol. 4, "Essay Third. On Taste," p. 318, includes taste as one of the "intellectual processes, which, by often passing through the mind, come at length to be carried on with a rapidity that eludes all our efforts to remark it; giving to many of our judgments, which are really the result of thought and reflection, the appearance of instantaneous and intuitive perceptions. The most remarkable instance [of these] . . . are commonly called the *acquired perceptions of sight* . . ."

P. 325, "The fact seems to be . . . 'the mind, when once it has felt the pleasure, has little inclination to retrace the steps by which it arrived at it.' It is owing to this, that Taste has been so generally ranked among our original faculties; and that so little attention has hitherto been given to the process by which it is formed."

44. *Ibid.*, p. 327.

45. *Ibid.*, p. 332: " 'The feeling', [Voltaire] observes, 'by which we distinguish beauties and defects in the arts, is prompt in its discernment, and anticipates reflection, like the sensations of the tongue and palate. Both kinds of Taste, too, enjoy, with a voluptuous satisfaction, what is good; and reject what is bad, with an emotion of disgust. Accordingly,' he adds, 'this metaphorical application of the word *taste,* is common to all known languages.' "

46. Date not traced, *OUN* 22–24.

47. Lessing, Gotthold Ephraim, *Laocoön. Nathan the Wise. Minna von Barnhelm,* William A. Steel, Ed., London, J. M. Dent, 1930: "The general distinguishing excellence of the Greek masterpieces in painting and sculpture Herr Winkelmann places in a noble simplicity. . . . in arrangement and in expression." And: "And if we now refer this to the Laocoön, the motive for which I am looking becomes evident. The master was striving after the highest beauty, under the given circumstances of bodily pain. This, in its full deforming violence, it was not possible to unite with that. . . . the aspect of pain excites discomfort without the beauty of the suffering subject changing this discomfort into the sweet feeling of compassion."

48. *Ibid.*

49. *Ibid.*: "[The artist] remains within the narrow range of a few designs, become familiar both to him and to everybody, and directs his inventive faculty merely to changes in the already known and to new combinations of old subjects. That, too, is actually the idea which the manuals of painting connect with the word *Invention* . . .

"In fact the poet has a great advantage who treats a well-known story and familiar characters."

50. In his copy of Abercrombie, *op. cit.*, pp. 199–203, Darwin wrote in the margins, "A man may wish to jump from a bridge to save another, but absolutely will not let him.—makes the muscles fail, & the heart sink—

Yes, but what determines his *consideration*—his own previous conduct—& what has determined that? & so on—*Hereditary* character & education—& chance (aspect of his will) circumstances.

Change of character possible from change of organization
What has given these *desires & conduct*
Then why·does not act of insanity give shame??

According to all this, ones disgust at villain <ought to be> is nothing more than disgust at some one under foul disease, & pity accompanies both. Pity ought to banish disgust. P→P For wickedness is no more a man's fault than bodily disease!! (Animals do persecute the sick as if were their fault). If this doctrine were believed—pretty world we should be in!—But it could not be believed excepting by intellectual people—if I believed it—it would make not one difference in my life, for I feel more virtue more happiness—

Believers would /will/ only marry good women & pay detail attention to education & so put their children in way of being happy. It is yet right to punish criminals for public good. All this delusion of free will would necessarily follow from man feeling power of action.

View no more unreasonable than that there should be *sick* & therefore unhappy men.

What humility this view teaches.

A man <reading> hearing bible by chance becomes good. This is effect of accident with his state of desire (neither by themselves sufficient) effect of birth & other accidents: may be congratulated, but deserves no credit.

When opposed desires are absolutely equal which is possibility, may free-will then decide—but it must be decided by habit or wish & these all originate as before."

51. *Ibid.* At the bottom of page 206 and 207, in his copy of Abercrombie, Darwin wrote, "A man may put himself in the way of above accidents. but desire to do so arises as before, & knowledge that the effect will be good, arise as before. education & mental disposition—

One feels how many actions, not determined by will, passion—when the motive power feeble & complicated & opposed we say free will (or chance) "

52. Darwin has reference to this trait of the wild, white cattle of Britain, e.g., the Chillingham Cattle.

53. Malthus, *op. cit.*

54. Darwin drew a heavy blue crayon line in the margin beside this paragraph.

55. This incident is also mentioned in *N* 63.

56. *The London and Westminster Review.* No. 65, for March 1840, pp. 139–162 (a review of seven of Coleridge's Works) .

57. *Ibid.*, p. 144: "We see no ground for believing that anything can be the object of our knowledge except our experience, and what can be inferred from our experience by the analogies of experience itself. . . ."

58. Date not traced, *OUN* 34–41.

59. Kirby, William, *On the Power, Wisdom, and Goodness of God, as Manifested in the Creation of Animals and in their History, Habits, and Instincts,* 2 vols., (*The Bridgewater Treatises*) , Pickering, London, 1835, Vol. 1, p. xli: " 'We have seen,' says [Lamarck], 'that the life which we remark in certain bodies, in some sort resembled nature, insomuch that it is not a being, but an order of things animated by movements; which also has its power, its faculties, and which exercises them necessarily while it exists.' "* (*Anim. sans Vertebr.* i. 321.)

60. Lamarck, *op. cit.*: ". . . we may include what essentially constitutes life in the following definition.

"*Life, in the parts of any body which possesses it, is an order and state of things which permit of organic movements; and these movements constituting active life result from the action of a stimulating cause which excites them.*"

61. Kirby, *op. cit.*, Vol. 1, pp. xli– xlii: "Speaking of the imponderable incoercible fluids, and specifying heat, electricity, the magnetic fluid, etc., to which he [Lamarck] is inclined to add light, he says, it is certain that without them, or certain of them, the phenomenon of life could not be produced in any body . . . [but] neither caloric nor electricity, though essential concomitants of life, form its essence."

62. *Ibid.*, xxiii–xxiv: "Body [Lamarck] observes, being essentially constituted of

cellular tissue, this tissue is in some sort the matrix, from the modification of which by the fluids put in motion by the stimulus of desire, membranes, fibres, vascular canals, and divers organs, gradually appear; and thus progressively new parts and organs are formed, and more and more perfect organizations produced; and thus, by consequence, in the lapse of ages, a monad becomes a man!!!"

63. *Ibid.*, pp. xxviii–xxix: "Every action of an intelligent individual, whether it be a movement or a thought . . . is necessarily preceded by a want of that which has power to excite such action [according to Lamarck]. This want felt immediately moves the internal sentiment, and in the same instant, that sentiment directs the disposable portion of the nervous fluid . . .' "

64. In this discussion Darwin, with his own interpretation, parallels a similar treatment in Kirby, *ibid.*, Vol. 1, pp. 150–151, eg., p. 150. "Lamarck indeed regards them [Infusoria] as having no volition, as taking their food by absorption like plants; as being without any mouth, or internal organ; in a word as transparent gelatinous masses, whose motions are determined not by their will, but by the action of the medium in which they move." P. 151, "Admitting that the observations of Spallanzani just stated record facts, it appears clearly to follow from them that these animals *have* volition, and therefore cannot properly be denominated *apathetic,* or insensible. The fact that they almost all have a mouth and a digestive system; many of them eyes, and some rudiments of a nervous one, implies a degree; more or less, of sensation . . ."

65. Kirby, *ibid.*, Vol. 2, p. 247: "That the same action should unfold such an infinite variety of forms in one case and instincts in the other is equally astounding and equally difficult to explain.—Compare the sunflower and the hive-bee, the compound flowers of the one, and the aggregate combs of the other . . ."

66. *Ibid.*, p. 246: ". . . as the most remarkable instincts of animals are those connected with the propagation of the species, so the analogue of these instincts in plants is the developement of these parts peculiarly connected with the production of the seed . . ."

Pp. 247–248: "Again, as all plants have their appropriate fruitification, so they have other peculiarities connected with their situation, nutriment, and mode of life, corresponding in some measure with these instincts that belong to other parts of an animal's economy. Some with a climbing or voluble stem, constantly turn one way, and some as constantly turn another. . . . others close their leaves in the night, and seem to go to sleep; others shew a remarkable degree of irritability when touched . . ."

67. Kirby, *ibid.*, Vol. 1, p. xxviii: "[Lamarck] admits [man] to be the most perfect of animals, but instead of a son of God, the root of his genealogical tree, according to him, is an animalcule, a creature without sense or voluntary motion, or internal or external organs . . . no wonder therefore that he considers his intellectual powers, not as indicating a spiritual substance derived from heaven though resident in his body, but merely as the result of his organization* (*N. Dict. D'Hist. Nat. xvi. Artic. Intelligence, 344. comp. Ibid. Artic. Idéa, 78, 80.) , and ascribes to him in the place of a soul, a certain *interior sentiment . . .*"

See also *B* 232, "The soul by consent of all is superadded . . ."

68. Kirby, *ibid.*, Vol. 1, pp. xxiv–xxv: "When indeed, one reads the above account [progression of molecules to monad to man] of the mode by which, according to [Lamarck's] hypothesis, the first vegetable and animal forms were produced, we can scarcely help thinking that we have before us a receipt for making the organized beings at the foot of the scale in either class—a mass of irritable matter formed by *attraction,* and a *repulsive* principle to introduce into it and form a cellular tissue, are the only ingredients necessary. Mix them, and you have an animal which begins to absorb fluid, and move about as a monad or a vibrio, multiplies itself by scissions or germes . . ."

69. *Ibid.*, p. xxix: ". . . Lamarck sees nothing in the universe but bodies, whence he confounds sensation with intellect."

70. *Ibid.*, p. xxvii: "Lamarck's great error, and that of many others of his compatriots, is materialism; he seems to have no faith in any thing but *body,* attributing

every thing to a physical, and scarcely any thing to a metaphysical cause." P. xxxiv: "From [Lamarck's] statements . . . he appears to admit the existence of a Deity . . ."

71. Pincher—pet dog.

72. Mark, Dr. Darwin's coachman at Shrewsbury.

73. Wynne: See p. 423: "Questions for Mr. Wynne."

74. Except for the insertions and the last paragraph, the document numbered 39–41 is not in Darwin's handwriting. Dr. Edward Manier of The University of Notre Dame Philosophy Department believes this document to have been written by Hensleigh Wedgwood. (Note: the emphases, the asterisk footnote, and the strikeouts, are Darwin's.)

75. Mackintosh, 1837, *op. cit.*, p. 113: "To this [the opinion that man as well as other animals prefer their own interest to every other object] Grotius answered, that even inferior animals under the powerful though transient impulse of parental love, prefer their young to their own safety or life; that gleams of compassion, and, he might have added, of gratitude and indignation, appear in the human infant long before the age of moral discipline; that man at the period of maturity is a social animal, who delights in the society of his fellow-creatures for its own sake, independently of the help and accommodation which it yields. . . ." Darwin has a marginal line beside this passage in his copy, now in the Cambridge University Library.

76. Darwin drew a double line in the margin beside the lines from "some force" to "should expect."

77. Darwin drew a marginal line beside this bracketed passage.

78. A marginal line beside the last two sentences.

79. A marginal line beside the lines from "conscience" to "tells."

80. A marginal line beside the lines from "One does not feel" to "power."

81. A marginal line beside the passage beginning "So we say" to "follow."

82. Mackintosh, 1837, *op. cit.*, p. 199: "But volitions and actions are not themselves the end, or last object in view, of any other desire or aversion. Nothing stands between the moral sentiments and their object. They are, as it were, in contact with the will . . . Conscience may forbid the will to contribute to the gratification of a desire. No *desire* ever forbids *will* to obey *conscience*." [Darwin's underlines.] P. 201, ". . . man becomes happier, more excellent, more estimable, more venerable, in proportion as conscience acquires a power of banishing malevolent passions . . ."

P. 198, "The truth seems to be, that the moral sentiments in their mature state, are *a class of feelings which have no other object but the mental dispositions leading to voluntary action, and the voluntary actions which flow from these dispositions.*" Darwin drew a line beside this passage, and made the following marginal notation, "How can cowardice, or avarice, or unfeelingness be said to be dispositions leading to action, yet conscience rebukes a man who allows another to drown without trying to save his life."

83. A marginal line by this sentence.

84. A marginal line by this sentence.

85. Mackintosh, 1837, *op. cit.*, n. p. 255: "A very ingenious man, Lord Kames, whose works had a great effect in rousing the mind of his contemporaries and countrymen, has indeed fancied that there is a 'hoarding instinct' in man and other animals. But such conclusions are not so much objects of confutation, as ludicrous proofs of the absurdity of the premises which lead to them."

86. *Ibid.*, p. 257: "It is easy to perceive how the complacency inspired by a benefit, may be transferred to a benefactor, thence to all beneficent beings and acts. The well-chosen instance of the nurse familiarly examplifies the manner in which the child transfers his complacency from the gratification of his senses to the cause of it; and thus learns an affection for her who is the source of his enjoyment. With this simple process concur, in the case of a tender nurse, and far more of a mother, a thousand acts of relief and endearment, of which the complacency is fixed on the person from whom they flow, and in some degree extended by association to all who resemble that person." Darwin has a double line beside this passage, and he wrote in the margin, "Common to animals hence Love of Place.—X! X Will not explain love of parent to child—except hereditary—"

87. Mackintosh, 1837, *op. cit.*, p. 152: "Reason, as reason, can never be a motive to action."

88. Probably *Eugenius; or, The Infidel Reclaimed*, 2 vols., Dawson, Stockport, London, 1830, Vol. 1, p. 116: "He [i.e., man] receives hourly additional means of cultivating virtue, and of maturing moral excellence. . . ." Pp. 145–146: "When we connect education with matters of most consequence . . . morality . . . I am persuaded it imperiously demands to be soberly, scrupulously, and sedulously conducted."

89. *Ibid.*, p. 261: "The sentiment of *moral approbation*, formed by association out of antecedent affections, may become so perfectly independent of them, that we are no longer conscious of the means by which it was formed, and never can in practice repeat, though we may in theory perceive, the process by which it was generated. It is in that mature and sound state of our nature that our emotions at the view of *Right* and *Wrong* are ascribed to *Conscience.*" Darwin drew a vertical marginal line beside this passage, and wrote, "rather instinctive" in the margin.

P. 262: "The pleasures (so called) of Imagination appear, at least in most cases, to originate in association. But it is not till the original cause of the gratification is obliterated from the mind, that they acquire their proper character. Order and proportion may be at first chosen for their convenience: it is not until they are admired for their own sake that they become objects of taste."

90. *Ibid.*, p. 194: "This natural supremacy [of man in nature] belongs to the faculty which surveys, approves, or disapproves the several affections of our minds and actions of our lives. As self-love is superior to the private passions, so conscience is superior to the whole man. Passion implies nothing but an inclination to follow it; and in that respect passions differ only in force. But no notion can be formed of the principle of reflection, or conscience, which does not comprehend judgment, direction, superintendency. 'Authority over all other principles of action is a constituent part of the idea of conscience, and cannot be separated from it.' " N.B.: The quotation marks were added by Darwin in pencil, and in the margin beside this statement he wrote, "if so, my theory goes.—in child one sees pain & pleasure struggling." (This passage is in the Section on Butler.)

91. *Ibid.*, p. 38: "According to [Mackintosh], the moral faculty consists of a class of desires and affections which have dispositions and volitions for their sole object . . . The moral sentiments are *in contact with the will* . . ." P. 36: "Man's soul at first, says Professor Sedgwick, is one unvaried blank, till it has received the impressions of external experience."

92. *Ibid.*, p. 333: "[Mr. Dugald Stewart] considers the appearance of moral sentiment at an early age, before the general tendency of actions could be ascertained, as a decisive objection to the origin of these sentiments in association,—an objection which assumes that, if utility be the criterion of morality, associations with utility must be the mode by which the moral sentiments are formed, which no skilful advocate of the theory of association will ever allow. That the main, if not sole, object of conscience is to govern our voluntary exertions, is manifest. But how could it perform this great function if it did not *impel the will?* and how could it have the latter effect as a mere act of reason, or indeed in any respect otherwise than as it is made up of *emotions,* by which alone its grand aim could in any degree be attained? Judgment and reason are therefore preparatory to conscience, not properly a part of it." Darwin has drawn a marginal line beside this passage, and has written in the margin, "can the instinct of bird building nest be said to imply will.—" And at the bottom of the page he wrote, "yet emotions are results—are trains of thought <firmly> long associated with action." N.B.: Darwin underlined "impel the will" and "emotions."

Pp. 376–377: "But it may still be reasonably asked, why these useful qualities [pursuit of truth and knowledge for their own sake, without regard for power or fame] are morally improved, and how they become capable of being combined with those public and disinterested sentiments which principally constitute conscience? The answer is, because they are entirely conversant with volitions and voluntary actions, and in that respect resemble the other constituents of conscience, with which they are thereby fitted to mingle and coalesce." Darwin drew a marginal line beside this passage, and wrote, "Nonsense—similar association may be made with actions, involuntary, as ——— [CD's blank] & etiquettes of society broken unconsciously.—" And beside the following passage Darwin has a marginal line and a large question mark and two large exclamation marks: "All those sentiments of which the final object is a state of the will, become thus intimately and inseparably blended; and of that perfect state of solution . . . the result is *Conscience*—the judge and arbiter of human conduct; which though it does not supercede *ordinary motives* of virtuous feelings and habits, which are the ordinary motives of good actions, yet exercises a lawful authority even over them, and ought to blend with them."

93. Eugenius, *op. cit.*, Vol. 2, pp. 25–26: "I indulged intense contemplation of the distant but flattering prospect, till my ideas grew vivid, my passions warm, and borne on fancy's soaring pinions, I winged my way thither. . . . When nearly examined, however, the entrancing forms, so lately inspiring ecstasy, vanished, and left me in the blackness of total midnight."

94. Mackintosh, 1837, *op. cit.*, p. 13: "The benevolent and family affections, and the desire of power, appear, then, to differ in some other way than in being modifications of the same elements; and, even if we choose . . . to call the latter class of principles selfish, the former must be arranged in a different group, which we cannot designate better than by calling them disinterested."

95. *Ibid.*, p. 14: "It is allowed on all sides that we have a conception of moral obligation; and the question is, Whether this conception can be resolved into some operation of the intellectual faculties, as the perception of general utility; or whether, on the contrary, it is incapable of being thus resolved, and must properly be ascribed to a separate faculty." The insertion is Darwin's.

96. Eugenius, *op. cit.*, Vol. 2, p. 63: ". . . they [i.e., the early Christian martyrs] underwent evil usage, not on account of crimes, but for well-doing; as people instrumental in promoting the peace and happiness of rational intelligence . . ." P. 214: "[Hume] endeavours, in his treatise on morals, to abolish the distinction between virtue and vice—to annihilate the sense of right and wrong, inherent in the constitution of things, innate in the conscience of each rational being . . ."

97. Mackintosh, 1837, *op. cit.*, p. 21: "[Paley] reduces moral obligation to two elements—external restraint, and the command of a superior. This attempt at an analysis of morality is singularly futile . . . external constraint annihilates the morality of the act, and the reference to a superior presupposes moral obligation . . . If Paley had stated his question . . . 'Why *ought* I to keep my word?' he would have had before him a problem more to the purpose of moral philosophy, and one to which his answer would have been palpably inapplicable."

98. *Ibid.*, p. 22: "Thus, as we separate the affections from the desires, we distinguish the moral sense, or conscience, from both. Butler, and Mackintosh with him, express the relation of conscience to the other principles of action, by ascribing to it a *supremacy*, or a right of command."

99. *Ibid.*, p. 240, "For it is certain that in many, nay in most cases of moral approbation, the adult man approves the action or disposition merely *as right*, and with a distinct consciousness that no process of sympathy intervenes between the approval and its object." Darwin made a marginal line beside this passage, and at the bottom of the page wrote, "My whole question with the breaking mere rule of etiquette."

100. *Ibid.*, pp. 251–252, in section on David Hartley, "[Mr. Gay] blames perhaps justly, that most ingenious man [Hutcheson], for assuming that these sentiments and affections are implanted, and partake of the nature of instincts . . ." "This precious mine may therefore be truly [be] said to have been opened by Hartley; for he who did such superabundant justice to the hints of Gay, would assuredly not have withheld the like tribute from Hutcheson, had he observed the happy expression of 'secondary passions' . . ."

Essay on Theology and Natural Selection

The document transcribed in this chapter is a set of notes jotted by
Darwin as he read selected passages from Macculloch's book *Proofs
and Illustrations of the Attributes of God*. In the notes Darwin states,
or alludes to, most of the postulates of his theories of Transmutation
and Natural Selection. If the paper, as evidence indicates, was in fact
written in the fall of 1838, then it has special historical importance.
The evidence, which is circumstantial, is (1) the paper has an 1838
watermark, (2) Malthus' postulates are alluded to (Darwin had read
Malthus beginning September 28, 1838), and (3) he cites Macculloch
throughout the essay, and he also cited Macculloch in the N notebook
(p. 35) in a passage written between October 30 and November 20.
The latter citation was almost certainly made after he had scanned
Macculloch's book.

In his *Autobiography* Darwin said he had determined, after having
grasped the idea of evolution through natural selection, that "to avoid
prejudice" he would for some time not write even the "briefest sketch"
of his theory. The jotted notes in this transcribed document may not
qualify as a "brief sketch," but they do constitute as complete a
résumé of his theories as exists in any other single, short document. No
such similar outline of his theories is in the transmutation notebooks,
even though he continued writing them through the year 1839. And
it wasn't until 1842, nearly four years after reading Malthus, that he
allowed himself the "satisfaction" of writing out the first acknowledged
"briefest sketch." Another interesting aspect of this early undated discourse is that in it Darwin tests the power of his own theory against
that of Providential Design in explaining the origin of special adaptations. He concludes that the Creationists' theory lacks predictive power,
and is also deficient in explaining many known sets of facts, such as, for
example, geographical patterns of species' dispersions, vestigial organs,

unnecessary variations, etc. To Darwin, the simple Malthusian postulates of checks and population pressures, and of Candollian struggles and wars, and of accidental and slight variations (the "grain of small advantages") were sufficient to explain even the most bizarre adaptations.

In a different parcel of the Darwin Manuscripts in the Cambridge University Library (in Box B) is a recently discovered four-page document which appears to have been originally a part of the same set of Macculloch notes transcribed in this Chapter. Darwin at some later date must have separated these newly found notes and put them with a collection of notes on Natural Selection in preparation for writing one of the early versions of the *Origin of Species*. This Box B paper is

202   REASON.

in their natural actions, we endeavour, as far as may be in our power, to supply those collateral circumstances, —by instructing him in the facts, truths, or motives; —by rousing his attention to their importance;—by impressing them upon him in their strongest characters, and by all such arguments and representations as we think calculated to fix the impression. All this we do under a conviction, that these causes have a certain, fixed, uniform, or necessary action, in regard to human volition and human conduct; and it is this conviction which encourages us to persevere in our attempts to bring the individual under their influence. If we had not this conviction, we should abandon the attempt as altogether hopeless; because we could have no ground on which to form any calculation, and no rules to guide us in our measures. Precisely in the same manner, when we find a chemical agent fail of the effect which we expect from it, we add it in larger quantity, or in a state of increased concentration, or at a higher temperature,—or with some other change of circumstances calculated to favour its action; and we persevere in these measures, under a conviction, that its action is perfectly uniform or necessary, and will take place whenever these circumstances have been provided for. On the same principle, we see how blame may attach to the intelligent agent in both cases, though the actions of the causes are uniform and necessary. Such is the action of chemical agents, but blame may attach to the chemist who has not provided them in the necessary cir-

FIRST TRUTHS.   203

cumstances, as to quantity, concentration, and temperature. Such is the action of moral causes,—but deep guilt may attach to the moral agent, who has been proof against their influence. There is guilt in ignorance, when knowledge was within his reach;—there is guilt in heedless inattention, when truths and motives of the highest interest claimed his serious consideration;— there is guilt in that corruption of his moral feelings which impedes the action of moral causes, because this has originated, in a great measure, in a course of vicious desires, and vicious conduct, by which the mind, familiarized with vice, has gradually lost sight of its malignity. During the whole of this course, also, the man felt that he was a free agent;—that he had power to pursue the course which he followed,—and that he had power to refrain from it. When a particular desire was first present to his mind, he had the power immediately to act with a view to its accomplishment; or he had the power to abstain from acting, and to direct his attention more fully to the various considerations and motives which were calculated to guide his determination. In acting as he did, he not only withheld his attention from those truths which were thus calculated to operate upon him as a moral being; but he did still more direct violence to an impulse within, which warned him that he was wandering from the path of rectitude. The state of moral feeling, which gradually results from this habitual violation of the indications of conscience, and this habitual neglect of the serious consideration of

Darwin marginalia.

of particular importance because in it Darwin discusses "the Malthusian rush for life." He points out that offspring which are by chance born with favorable adaptations will if circumstances permit have a greater probability of survival and therefore will have a reproductive advantage. He says that in ten thousand years the new race "will get the upper hand, though continually dragged back to old type by intermarrying with ordinary race." The manuscript of the *Essay on Theology and Natural Selection* is in the Cambridge University Library.

1   Macculloch. Attrib of Deity. Vol I: p. 280. adduces provisions of seeds for transportation through the air.— ((it will be better always to refer to the author if I use these facts)) cocoa nut by water /fucus for adhesion/.[1]—as example of design.—perhaps they are so.—but the coral rock might have been uninhabited as the alpine pinnacles.[2]—One thing must be admitted, there would not be these plants, if there was not some provision for transportation:—But I do not want to deny laws.—the whole universe is full of adaptations.—but these are, I believe, only direct consequences of still higher laws.—I do not /then/ believe, the [illegible] of any one seed (all have not it) was DIRECTLY *created* for transportation. it follows from some more general law.—[that the laws of propagation were created with reference to successive developement I admit, but the admission is probably from ignorance.][CD]

Who would even have thought that the intestine of a thrush was means sufficient to ensure propagation of misseltoe?— |

2   do [ditto] p. 284. it is hard on my theory of grain of small advantages thus to explain the curling of the valves of the broom.[3]—or the springing of other seeds.—But are we certain that these are necessary adaptations.—may they not be accidental? we have good reason to know that they would not be detrimental accidents, & domesticated variations show us accidents may become hereditary [produce some peculiarity in seed vessel][CD] if man takes care they are not detrimental.—N.B. One limit to the transmission of abortive organs will be as long as they are not detrimental.—

3   p. 285 the seed-pod of a desert plant (anastatica) is rolled along & splits when it comes to a damp place.[4]— ((Kohlreuter mentions some hybrid, whose flower great tendency to break off)) |

p. 292. Mac. has long rigmarole about plants being created to arrest mud etc at deltas.[5]—now my theory makes all organic beings perfectly adapted to all situations where in accordance to certain laws they can live.—Hence the mistake they are created for them. If we once venture to say plants created to <arrest> prevent the valuable soil in its seaward course, we sink into such contemptable quiries as why should the earth be drifted, why should plants require earth, why not created to live on alpine pinnacle? if we were to presume that God /created plant to/ arrest earth, (like a Dutchman plants them to stop the moving sand) we lower the creator to the standard of one of his weak creations.—All such facts are merely relations of one general law. the plants were no more created to arrest the earth, than the earth revolves to form rain to wash down earth from the mountains upheaved by volcanic forces, for these marsh plants. All flow from some grand & simple laws.— |

4   p. 308 ((Study Cuviers Anatomie Comparé)) [6] Traces the gradation of skeleton in Vertebrates[7] & constantly alludes (& at p. 312) to the abortive bones.[8] He explains it by saying "It is the determination to adhere to a plan once adopted: & it is from these very circumstances, that we become satisfied respecting an original thought, or design, pursued to its utmost exhaustion, & till it must be abandoned for another."[9]—The <design> determiner of a God-head.—the designs of an omnipotent creator, exhausted & abandoned. Such is man's philosophy, when he argues about his Creator!

p. 309. says the ribs in Draco support the flying membrane?!![10]—that the phalanges have separate movements in the Holocentrus ruber (a fish) [11]

Man has abortive muscles to his ears.—p. 313—many other good cases.—p. do <Mac. remarks all mammifers originally land animals.> |

5   p. 314 Mac. remarks all <land> mammifers animals originally terrestrial.[12]—for we find even in Cetaceae traces of hind extremities.—How are we to explain this.—Did reptiles first inhabit seas.—Were they then killed out /by the intense cold/, & did mammifers than take their place? Would they not first occupy the Poles? Is this origin of the Polar attributes of the Cetaceae.—How came Bats also? before birds? They are ancient.— Are Cetaceae found in Paris Basin?—

N.B. The explanation of types of structure in classes—as resulting from the *will* of the deity, to create animals on certain plans,—

is no explanation—*it has not the character of a physical law* /& is therefore utterly useless.—it foretells nothing/ because we know nothing of the will of the Deity, how it acts & whether constant or inconstant like that of man.—the cause given we know not the effect |

6   p. 412 Macculloch explains the shortness of life (peculiar to each species) owing to the *growing size of the world?*[13] & the physical changes it was to undergo /animals feeding on each other etc etc/. —There are reasons /causing death to some, etc, etc/ just as liability to accidents & any other cause.— (& my theory) [ALL PARTS OF THE GREAT SYSTEM. C.D.][CD] [All this does not explain *death,* but *reproduction*][CD] though such a scheme would require constant miracles.—[14]
p. 420 thinks the great fecundity of *germs* is to afford support to other beings.[15]—*true* (& the doctrines of checks & my theory) |

7   Macculloch. Attrib. Vol. I

p. 330. Mentions the many cases as in Papilionaceous flower, where such case seems to be taken that the anthers should not be exposed to weather.[16]—this is against my theory of frequent intermarriage.—A plant is in the same predicament as a group of bisexual animals living in the borders of a country favourable to change.—

It might be concluded that Plants would be subject to extreme variation as long as crossing with other varieties was prevented.

Do races of peas become intermixed & gardener have hybrid seedlings

p. 333 Macculloch brings forward the impregnation of Dioecious Plants by foreign agency—as insects, as wonderful case of adaptation![17] There would not have been any Dioecious plants, had there been no insects. The right inference is, there were insects (?when were Pollen formed?) as soon as Dioecious Plants were formed. |

8   Macculloch says life forms a broken, recurrent series, whilst the habitation /or world/ simple series.—My theory shows life equally simple series, & therefore trace of beginning in organic world.—

Macculloch. Attrib of Deity Vol. I, p. 232. gives Woodpecker as instance of beautiful adaptation.—& then Chamelion, which feeding on same food, differs in every respect, except in quick movement. (Sliminess instead of barbs) [18]—In all these cases it should be remembered, that animals could not exist without these adaptations.—fossil forms show such losses.—Consider *ground* Woodpecker stiff tailed cormorant: pain & disease in world & yet talk of perfection |

9   Get instances of adaptations in varieties.—greyhound to hare.—waterdog, hair to water—bulldog to bulls,—primrose to <open fields> banks—cowslip to <banks> (fields) —these are adaptations just as much as Woodpecker.—only we here see means—but not in the other.

All Bridgewater Treatises are reduced simply statements of productiveness, & laws of adaptation.

p. 234 The non-absorbing Camel's stomach a puzzler—[19]

p. do. says *inconvenience* would have arisen had /some/ insects not been provided with proboscis, /as bee & butterfly/.[20] inconvenience! *extinction,* utter *extinction!* let him study Malthus & Decandoelle.— |

10   The Final cause of innumerable eggs is explained by Malthus.—[is it anomaly in me to talk of Final causes: consider this?][CD] consider these barren Virgins

p. 235. talks of the long spinous processes in Giraffe etc as adaptation to long necks[21]—why they may as well say /long/ neck is adapted to long necks.—

p. 236. Marsupial bones especial adaptation, to young—good God & yet [illegible] have them.[22] What trash

p. 237 Gives as summary of adaptations horny point to chickens beak, to break egg shells[23]—why chicken could not have lived had it not been so.—let egg shells grow harder. so must those with weak beaks be sifted away.— |

p?   In the Mollusca /Bees/ the nervous system is endowed with the knowledge of trying a hundred schemes of structure, in the *course of ages* /step by step/.—in man, the nervous system, gains that knowledge, before hand, & can in idea (with consciousness) form these schemes.—

I see no reason why structure of brain should not be born, with tendency to make animal perform some action.—as well as gain it by habit.—New theory of instinct, returning to Kirby's[24] view.— |

11   & the species, like 10,000 others permits, & who will dare to say that this is an infringement on the wisdom on [of] Providence, when /whole/ rocks [illegible] very mountains are formed of such dead & extinct forms.— ((the excuviae of the dead & extinct))

The analogy between the worker of art /or intellect/ such as hinge
& hinge of shell, work of laws of organization is remarkable—what
is intellect, but organization, with mysterious consciousness superadded. This is similar idea to cells of bees, corresponding to [four
illegible words]—brain making structure, instead of parts of body.
—now we know what instinct is—consider this

I look at every adaptation, as the surviving one of ten thousand
trials.—each step being perfect ((or nearly so (except in 1st) although having hereditary organization. Man could exist without
mammae.)) to the then existing conditions.—An adaptation made
by intellect this process is shortened, but yet analogous, no savage
ever made a perfect hinge.—reason, & not death rejects the imper-
fect attempts. |

Notes by Paul H. Barrett

1. Macculloch, *op. cit.*, Vol. 1, p. 280: "[The seeds of fuci] are surrounded by a
mucilage which water cannot dissolve, and which enables them to adhere to whatever solid body they touch. . . ."

2. *Ibid.*, p. 279: "If the floating of seeds through water is a contrivance which,
like the action of the winds, appears too much akin to what we carelessly term
accident, to deserve notice, yet thus chieny are the naked coral rocks of the great
Pacific Ocean clothed with vegetation, and rendered fit for the habitation of man.
Are we entitled to give the name of accident to that cause, or combination of causes,
by which so great an end is produced—even though metaphysics, and religion
equally, did not show that there can be no accident to the Creator and Governor of
all things? The buoyancy of a cocoa nut . . . can be no accident. . . ."

3. *Ibid.*, p. 285: "In broom, the crackling of which in a hot day is familiar, each
valve recoils in a spiral direction when the detent yields. . . ."

4. *Ibid.*, p. 285: "Thus also is it with the rose of Jer̲̲ː (anastatica) , where
the seed-vessel is rolled along the sands by the winds, until, meeting with a moist
spot, it opens and parts with its seeds in that only place amid the parched plain
where provision has been made for their vegetation."

5. *Ibid.*, p. 291: ". . . who can doubt that the singularly long, powerful, and
prolific root of the latter [bulrush and common reed], was constructed for the very
purpose of consolidating the earth. . . ."

6. *Ibid.*, p. 307: "[Comparative anatomy] proves, not merely design, but One
intelligent and designing Creator, so does it exclude all systems of chance; the very
variations marking that steady pursuit of a single primary object, which no multiplicity of designs could have equally proved to those who have adopted this scheme
of Creation." (See also Cuvier, G., *Leçons d'anatomie comparée*, par Duméril, 5
vols., Baudouin, Paris, 1800. 1805.)

7. *Ibid.*, pp. 315–316: "But I must quit details which I have not space to pursue any further, to point out some familiar examples of what are termed gradations in the animal kingdom; or cases of intermixed structure, or of transition, such
as that just noticed from the quadruped to the whale."

8. *Ibid.*, p. 312: "Even in some serpents, the rudiments of the pelvis have been
traced, though apparently quite useless to them. . . ."

9. *Ibid.*, pp. 306–307.

10. *Ibid.*, p. 309: ". . . while in the serpents they [the main ribs] may almost be said to become feet; and in the flying lizards some of them are extended straight, so as to be the base of wings."

11. *Ibid.*, p. 309: ". . . in the Holocentrus ruber, the rows of phalanges become as independent as fingers. . . . Such variations as this render the whole progress of the design more uniform. . . ."

12. *Ibid.*, p. 314: "The unexpected nature of the Cetaceous fishes seems also best explained on the same principles: it is the terminal point of a plan commencing in the quadrupeds as land animals, and gradually traced to a perpetual residence in the water, through intermediate stages of construction or variation."

13. *Ibid.*, pp. 412–414: ". . . the mere enlargement of the earth is not all . . . it has differed in climates and soils. . . . Species beyond number were originally created to occupy every climate and every soil, every conceivable point of the world. . . . The constitutions and the inclinations of the animals must be altered to meet these changes . . . it could not have been occupied by an original and undying creation of animals, without some further provision . . . nothing but continued new creations, or a principle of reproduction, could have sufficed to fill the new blanks, or meet the varying changes of the earth's surface. . . . And while this demands the constant interposition of God, or a providence, in the most rigid sense of that term, that is one of the views which are held in particular disesteem by all those who would thus amend the order of nature. All hypotheses of this kind are connected with a system of general laws and non-interference: so difficult is it to be consistent, in remonstrating with the plans of the Deity, and in inventing systems of that philosophy so often stigmatized by a term which I suppress whenever that is possible" (i.e., Transmutation).

14. *Ibid.*, p. 415: "He must have created new species, equally immortal, to fill those new blanks, or He must have empowered the original species to increase in exact proportion and at the exact times required . . . nothing less than miraculous interposition could have prevented them from being destroyed by accidents."

15. *Ibid.*, p. 420: "The germs of plants are produced in myriads, without any design that they should grow into representatives of the original: this excessive fecundity is designed for the support of animals, and the mortality of the former is the life of the latter . . . it is from the mortality of one species that another receives the means of existence."

P. 421: "It is the same principle, too, which is now producing new islands, new continents; extending new territories for new races of life, and, under one great scheme of joint life and mortality, laying the preparations for an endless succession of new lives under new forms."

16. *Ibid.*, p. 329: "The anthera was . . . to be protected from rain and wind, lest its pollen should be lost. . . ."

P. 330: ". . . the peculiar forms of flowers seem directed to this sole end. . . . In papilionaceous flowers, the securities are multiplied beyond all apparent need or use."

17. *Ibid.*, p. 331: ". . . the meeting of the pollen and the stigma is the ultimate end of all this contrivance. . . ."

P. 333: "Insects are here [in the case of dioecious plants] the great agents. . . ."

18. *Ibid.*, p. 232: ". . . [the tongue of the woodpecker] is a sort of spear provided with barbs . . . and . . . the tail is rendered stiff, so as to be an assistant in climbing. . . ."

P. 232: "Similar food . . . was destined for the chameleon . . . and . . . the motion of the tongue has the rapidity of lightning. . . ."

19. *Ibid.*, p. 234: "The stomach of the camel offers another of those special contrivances, where the purpose, and the means of attaining that, are so perfectly adapted, that the design has been universally admitted. It was created to live in a land of little water. . . . This contrivance consists in certain appendages to one of the stomachs. . . . But it is ordered that the water receptacles of the camel shall not be absorbent . . . and thus is the perfection of this design evinced."

20. *Ibid.*, p. 234: "In the insect races, there is a very general case, where in-

convenience or evil would have followed . . . had there not been an analogous spec-ial contrivance made. . . . I allude to those insects in which the mouth is a pro-boscis, as in the bee and the butterfly."

21. *Ibid.*, p. 235: ". . . the long spinal processes of the withers, or upper dorsal vertebrae, in the animals of long necks, like the camelopard, where the purpose of giving a lever to the muscles is apparent. . . ."

22. *Ibid.*, p. 236: "The marsupial bones in the opossum race offer another in-stance of the same nature [of a special invention] . . . it being granted that the young required the protection which the pouch affords, the invention is per-fect. . . ."

23. *Ibid.*, p. 237: "I allude to the horny point on the beak of a chicken, with which it is supplied for the purpose of breaking the shell to procure its release, and falling off a few days after the birth."

24. See Kirby, *op. cit.*

Questions for M[r] Wynne[1]

In 1839 Darwin began to circulate a printed list of "Questions About the Breeding of Animals." A few copies have recently been found and published, together with the answers given by the animal breeders to whom they were sent.* The list was seven pages long with ample room for writing in the answers. The earliest definitely dated reply was written on May 6, 1839, by R. S. Ford, a farmer in the Staffordshire village of Swynnerton. (This was not far from Maer, the home of Darwin's uncle, Josiah Wedgwood, and of Darwin's wife, his cousin Emma Wedgwood.)

Apparently, before he had the questions printed up, Darwin drafted an earlier list, which we reproduce below. While this manuscript is undated, it is logical to suppose that Darwin would not have used a handwritten version if he had the printed one available. There are many differences between the two sets of questions. The questions in the printed list are very general, fully written out, and carefully explained—as would be necessary if they were to be answered in writing by a variety of respondents. The questions for Mr. Wynne are briefer and quite telegraphic in style, almost like an outline for an interview Darwin might have been planning. At a few points the questions turn into notes about his own ideas.

We have not identified Mr. Wynne or found his answers.

---

1 Transcribed from a copy in Darwin's handwriting, now in the Cambridge University Library.
* See R. B. Freeman and P. J. Gautrey, "Darwin's Questions about the Breeding of Animals, with a Note on Queries about Expression," *Journal of the Society for the Bibliography of Natural History*, Vol. 5, 1969, 220–225.

[1] Are offspring like fathers or mothers? How are real *nipples?*

Is a peculiarity which has long been in blood more easily transmitted, than a newly acquired one?

Does peculiarity [adhere?] [when?][2] geometrical progression is proportion to no of generations

When old variety is crossed with new; or natural one with very artificial one as shepherd dog with Italian Greyhound do offspring partake more of the natural, than artificial kind?—

When wild animal crossed with tame does offspring favour the former. fox with dog?—

Is a breed of half bred animals more subject to variation, than either parent stock? is unusual care required to keep breed constant

Superfetation cases of?

Idea of beauty in animals: do females prefer certain males? or vice versa when in flock.—

When beauty /strong or puny/ unhealthy animals /or men/ crossed do offspring partake more of former or latter?—Effects of habit on form.—in men as in trades.

Could you get racehorse from carthorse without training?

If horses temper soured, would be handed down.— |

[2] if temper cowed in horse or dog or cock. hereditary.—

Case of Malay fowls—habits?

Cross of Chinese pigs, are they intermediate in form. as in dogs.—does Mr Wynne believe in dogs

Cases of hereditary monsters? of accidental mutilations being hereditary.

Case of heterogeneous offspring, in fowls, pidgeons, rabbits.—if race horse & cart horse be crossed, will offspring be constant.—

Effects of crossing stocks with different constitutions?

If bull-dog be crossed with greyhound are they as prolific as rather nearer breed—/But/ Are the mongrels prolific.

Breeding in & in Infertility & loss of passion?? in Male?

What would effect be of own brother & sister taken to one country, one pair to other & made different. Would not the cousins cross. No be-

---

2 These two words difficult to decipher.

cause both would be bred in & in—[illegible] the male [illegible] to pick out

opposed animals, i.e. animals which have each acquired peculiarities <Horse> Is there not some strange fact about twin calfs one being *neutral.* how is it with animals
are sexes always same in twins
why.

About sporting in pack of Hounds: how much selection??[3]

Are all or some only of cross-bred animals more prolific??[4]

---

[3] Written vertically in left-hand margin of page one.
[4] Scrawled across upper part of page one.

PART II

Selections from
Previously Published Writings
of Charles Darwin

From the Beagle Diary

Commentary by Howard E. Gruber

Darwin's interest in man was the least formalized and therefore the least visible of his concerns as a naturalist during the voyage of the *Beagle*. In the fields of botany, zoology, and geology he kept specialized notes and sent home many boxes of specimens, enough to make a small museum. His efforts eventuated in five volumes of zoology and three of geology, all based on his own explorations and collections. On man there was no such definite product. But running through the several stages of the narrative of the voyage one sees Darwin's steady interest in man in all his varieties, in the relation between man and other animals, and in the effects of civilization.

After the voyage Darwin compiled his famous *Journal of Researches into the Geology and Natural History of the Various Countries Visited by H.M.S. Beagle, Under the Command of Captain Fitz-Roy, R.N. from 1832 to 1836*, often referred to as the "Voyage of the *Beagle*." Among the several kinds of notes he used in writing the *Journal*, his main source was his *Diary*, written between 1831 and 1836. Kept as a running record during the entire voyage, the *Diary* gives the flavor of Darwin's thoughts on many subjects not included in his scientific notes, such as the beauties of scenery, the fortunes of war, the foibles of man, the violence of conquest in the European settlement of South America, the feel of an overland journey on horseback, and Darwin's reflections on man's place in nature. We have excerpted a few passages from the *Diary*, to give the reader a sense of Darwin's early and continuing attention to the study of man.[1]

---

[1] We quote from *Charles Darwin's Diary of the Voyage of H.M.S. "Beagle,"* edited from the manuscript by Nora Barlow, Cambridge: University Press, 1934. This edition is a verbatim, entirely faithful transcription of the manuscript kept at Darwin's old home in Down.

On Primitive Man

Perhaps no single experience ever impressed Darwin more than his contacts with primitive human beings. For him, these observations were cast in a special light by the fact that he had first known as shipmates three Indians from Tierra del Fuego who were being returned from a sojourn in England, where they had been taken to learn civilized ways.

December 16, 1832: "We made the coast of Tierra del Fuego a little to the South of Cape St. Sebastian, & then altering our course ran along a few miles from the shore . . . 17th . . . We kept close to the Fuegian shore; the outline of the rugged inhospitable Staten land was visible amidst the clouds. In the afternoon we anchored in the bay of Good Success; here we intend staying some days. In doubling the Northern entrance, a party of Fuegians were watching us; they were perched on a wild peak overhanging the sea & surrounded by wood. As we passed by they all sprang up & waving their cloaks of skins sent forth a loud sonorous shout; this they continued for a long time. . . .

"December 18th. The Captain sent a boat with a large party of officers to communicate with the Fuegians. As soon as the boat came within hail, one of the four men who advanced to receive us began to shout most vehemently, & at the same time pointed out a good landing place. The women & children had all disappeared. When we landed the party looked rather alarmed, but continued talking & making gestures with great rapidity. It was without exception the most curious & interesting spectacle I ever beheld. I would not have believed how entire the difference between savage & civilized man is. It is greater than between a wild & domesticated animal, in as much as in man there is greater power of improvement. . . .

"January 20th. We began to enter to day the part of the country which is thickly inhabited. . . . I shall never forget how savage & wild one group was. Four or five men suddenly appeared on a cliff near to us; they were absolutely naked & with long streaming hair. Springing from the ground & waving their arms around their heads, they sent forth most hideous yells. Their appearance was so strange, that it was scarcely like that of earthly inhabitants."

The time came to put ashore the returning Fuegians who had been educated in England, Jemmy Button, York Minster, and Fuegia Basket.

February 6, 1833: ". . . It was quite melancholy leaving our Fuegians amongst their barbarous countrymen. There was one comfort; they appeared to have no personal fears. But, in contradiction of

what has often been stated, 3 years has been sufficient to change
savages into as far as habits go, complete & voluntary Europeans. York,
who was a full grown man & with a strong violent mind, will I am
certain in every respect live as far as his means go, like an Englishman.
Poor Jemmy looked rather disconsolate & certainly would have liked to
return with us; he said 'they were all very bad men, no "sabe" nothing.'
Jemmy's own brother had been stealing from him; as Jemmy said,
'what fashion do you call that.' I am afraid whatever other ends this
excursion to England produces, it will not be conducive to their happiness. They have far too much sense not to see the vast superiority of
civilized over uncivilized habits, yet I am afraid to the latter they must
return."

As time went on, Darwin had opportunities to compare the very
primitive Fuegians with other Indians. Encountering a group of Pata-
gonian Indians, he wrote:

August 14, 1833: "The men are a tall exceedingly fine race; yet it
is easy to see the same countenance rendered hideous by the cold, want
of food, & less civilization, in the Fuegian savage. Some authors in de-
fining the primary races of man, have separated these two classes of
Indians, but I cannot think this is correct."

At Chiloe, an island on the west coast of South America, off
Chile, he wrote, on November 26, 1834:

"When landing on a point to take observations, we saw a family
of pure Indian extraction; the father was singularly like to York
Minster; some of the younger boys, with their ruddy complexions,
might be mistaken for Pampas Indians. Everything I have seen con-
vinces me of the close connection of the different tribes, who yet speak
quite distinct languages. This party could muster but little Spanish &
talked to each other in their own dialect. It is a pleasant thing in any
case to see the aboriginal inhabitants, advanced to the same degree of
civilization, however low that may be, which their white conquerors
have attained."

Summing up his experiences, at the close of the voyage he wrote in
late September 1836, on the way home:

"Of individual objects, perhaps no one is more sure to create astonishment, than the first sight in his native haunt, of a real bar-
barian,—of man in his lowest & most savage state. One's mind hurries
back over past centuries, & then asks, could our progenitors be such as
these? Men,—whose very signs & expressions are less intelligible to us
than those of the domesticated animals; who do not possess the instinct of those animals, nor yet appear to boast of human reason, or at
least of arts consequent on that reason. I do not believe it is possible to
describe or paint the difference of savage & civilized man. It is the
difference between a wild & tame animal: & part of the interest in be-

holding a savage is the same which would lead every one to desire to see the lion in his desert, the tiger tearing his prey in the jungle, the rhinoceros on the wide plain, or the hippopotamus wallowing in the mud of some African river."

On Civilization

Darwin took a qualified view of the virtues of civilization. He traveled in South America during a period of revolutionary wars against Spain and genocidal campaigns of extermination of Indians. He did not always see white civilized man at his English best, nor did the Indians of South America always come off poorly in the comparisons Darwin made.

September 4–7, 1833, in Argentina: "The Indians are now so terrified, that they offer no resistance in body, but each escapes as well as he can, neglecting even his wife & children. The soldiers pursue & sabre every man. Like wild animals, however, they fight to the last instant. . . . This is a dark picture; but how much more shocking is the unquestionable fact, that all the women who appear above twenty years old are massacred in cold blood. I ventured to hint that this appeared rather inhuman. He answered me, 'Why, what can be done? they breed so!' Every one here is fully convinced that this is the justest war, because it is against Barbarians. Who would believe in this age in a Christian civilized country that such atrocities were committed? The children of the Indians are saved, to be sold or given away as a kind of slave, for as long a time as the owner can deceive them.

". . . If this warfare is successful, that is if all the Indians are butchered, a grand extent of country will be gained for the production of cattle: & the vallies of the R. Negro, Colorado, Sauce will be most productive in corn. The country will be in the hands of white Gaucho savages instead of copper coloured Indians. The former being a little superior in civilization, as they are inferior in every moral virtue."

Darwin took a brighter view of British than of Spanish imperialism. Although a patriotic young Englishman, he was not entirely oblivious to the faults of his countrymen.

January 12, 1836, Sydney, Australia: "At last we anchored within Sydney Cove; we found the little basin containing many large ships & surrounded by Warehouses. In the evening I walked through the town & returned full of admiration at the whole scene. It is a most magnificent testimony to the power of the British nation: here in a less promising country, scores of years have effected many times more than centuries in South America. My first feeling was to congratulate myself that I was born an Englishman. Upon seeing more of the town on other days, perhaps it fell a little in my estimation; but yet it is a good town. . . ."

February 5, 1836, Hobart Town. "The Aboriginal blacks are all removed & kept (in reality as prisoners) in a Promontory, the neck of which is guarded. I believe it was not possible to avoid this cruel step; although without doubt the misconduct of the Whites first led to the Necessity."[2]

Man and Other Animals

A feeling for man as an animal marked Darwin's thinking from a very early moment.

November 14, 1833, near Monte Video: ". . . I was amused by seeing the dexterity with which some Peons crossed over the rivers. As soon as the horse is out of its depth, the man slips backwards & seizing the tail is towed across: on the other side he pulls himself on again. A naked man on a naked horse is a very fine spectacle; I had no idea how well the two animals suited each other: as the Peons were galloping about they reminded me of the Elgin marbles."

Musing on the easy availability of food in some tropical places, November 18, 1835, Tahiti: ". . . I could not look on the surrounding plants without wonder . . . the fruits which served for food in many ways, lay in heaps decaying on the ground. . . . The little stream, besides its cool water, produced also eels & cray-fish. I did indeed admire this scene, when I compared it with an uncultivated one in the temperate zone. I felt the force of the observation that man, at least savage man, with his reasoning powers only partly developed, is the child of the Tropics. (One cannot however say that one is more natural than the other; if an animal exerts its instinct to procure food, the law of Nature clearly points out that man should exert his reason & cultivate the ground.) "[3]

---

2 This description seems excessively kind to the British colonists. In the published *Journal,* Darwin added some remarks on the hunting down of the aborigines: "Although numbers of natives were shot and taken prisoners in the skirmishing which was going on at intervals for several years, nothing seems fully to have impressed them with the idea of our overwhelming power, until the whole island, in 1830, was put under martial law, and by proclamation the whole population desired to assist in one great attempt to secure the entire race. The plan adopted was nearly similar to that of the great hunting-matches in India: a line reaching across the island was formed, with the intention of driving the natives into a *cul-de-sac* on Tasman's peninsula." *Journal of Researches,* 1839, pp. 533–534.

3 The passage in parentheses was deleted from the published *Journal.*

Natural Selection

Darwin wrote little during the voyage suggesting any direct interest in evolution. There are nevertheless a few interesting passages in the *Diary* bearing on the process of selection. After visting a gold mine he dwelt briefly on a mechanical process having some characteristics analogous to selection. There are, of course, numerous passages dealing with extinction. Some of these, based on his fossil finds, treat extinction as a process remote in time; others, especially those dealing with human populations, treat it as an ongoing process. Among the most striking of these passages is his comment on January 12, 1836, showing an explicitly Malthusian awareness of the relation between variations in population and in food supply.

September 13, 1834, Chile: "When the ore is brought to the Mill it is ground into an impalpable powder; the process of washing takes away the lighter particles & amalgamation at last secures all the gold dust. The washing when described sounds a very simple process: but it is at the same time beautiful to see how the exact adaptation of the current of water to the Specific Gravity of the gold so easily separates it from its matrix. It is curious how the minute particles of gold become scattered about, & not corroding, at last accumulate even in the least likely spots."

January 4, 1835, Cape Tres Montes, Chile: "The entire absence of all Indians amongst these islands is a complete puzzle. That they formerly lived here is certain, & some even within a hundred years; I do not think they could migrate anywhere, & indeed, what could their temptation be? For we here see the great abundance of the Indians' highest luxury;—seal's flesh. I should suppose the tribe has become extinct; one step to the final extermination of the Indian race in S. America."

January 12, 1836, New South Wales: "At Sunset by good fortune a party of a score of the Aboriginal Blacks passed by, each carrying, in their accustomed manner, a bundle of spears & other weapons. By giving a leading young man a shilling they were easily detained & they threw their spears for my amusement. They were all partly clothed & several could speak a little English; their countenances were good-humoured & pleasant & they appeared far from such utterly degraded beings as usually represented. In their own arts they are admirable; a cap being fixed at thirty yards distance, they transfixed it with the spear delivered by the throwing stick, with the rapidity of an arrow from the bow of a practised Archer; in tracking animals & men they show most wonderful sagacity & I heard many of their remarks, which manifested considerable acuteness. They will not, however, cultivate the ground, or even take the trouble of keeping flocks of sheep which have been offered them; or build houses & remain stationary. Nevertheless, they appear to me to stand some few degrees higher in civilization, or more correctly, a few lower in barbarism, than the Fuegians.

"It is very curious thus to see in the midst of a civilized people, a set of harmless savages wandering about without knowing where they will sleep, & gaining their livelihood by hunting in the woods. Their numbers have rapidly decreased; during my whole ride with the exception of some boys brought up in the houses, I saw only one other party. These were rather more numerous & not so well clothed. I should have mentioned that in addition to their state of independence of the Whites, the different tribes go to war. In an engagement which took place lately the parties, very singularly, chose the centre of the village of Bathhurst as the place of engagement; the conquered party took refuge in the Barracks. The decrease in numbers must be owing to the drinking of Spirits, the European diseases, even the milder ones of which such as the Measles, are very destructive, & the gradual extinction of the wild animals. It is said, that from the wandering life of these people, great numbers of their children die in very early infancy. When the difficulty in procuring food is increased, of course the population must be repressed in a manner almost instantaneous as compared to what takes place in civilized life, where the father may add to his labor without destroying his offspring."

Slavery and Imperialism

July 3, 1832, Rio de Janeiro: "The state of the enormous slave population must interest everyone who enters the Brazils. Passing along the streets it is curious to observe the number of tribes which may be known by the different ornaments cut in the skin & the various expressions. From this [number of tribes] results the safety of the country. The slaves must communicate amongst themselves in Portugeese & are not in consequence united. I cannot help believing they will ultimately be the rulers. I judge of it from their numbers, from their fine athletic figures (especially contrasted with the Brazilians) proving they are in a congenial climate, & from clearly seeing their intellects have been much underrated; they are the efficient workmen in all the necessary trades. If the free blacks increase in numbers (as they must) & become discontented at not being equal to white men, the epoch of the general liberation would not be far distant. . . . I hope the day will come when they will assert their own rights & forget to avenge their wrongs."

September 25, 1836, on the voyage home: "From seeing the present state, it is impossible not to look forward with high expectation to the future progress of nearly an entire hemisphere. The march of improvement, consequent on the introduction of Christianity, through the South Sea, probably stands by itself on the records of the world. It is the more striking when we remember that but sixty years since, Cook, whose most excellent judgment none will dispute, could foresee no prospect of such change. Yet these changes have now been effected by the philanthropic spirit of the English nation."

Religion and Nature

Darwin became a confirmed agnostic, probably sometime about 1840. During the *Beagle* voyage his religious views must have been in flux. He began the voyage aspiring to return to a "quiet parsonage" in England; he ended it a single-minded scientist. His devotion to the Church of England may have faded early in the voyage. The *Diary* is not deeply imbued with conventional religious feeling. There are passages which criticize deeds done in the name of Christianity, passages objectively comparing different sects, and still others which stress the unity of mankind in sharing some sense of awe and reverence toward nature; there is at least one passage in which religion and patriotism are confused with each other.

January 29, 1832, St. Jago: "Divine service was performed on board; it is the first time I have seen it: it is a striking scene & the extreme attention of the men renders it much more imposing than I had expected. Everything on board on Sunday is most delightfully clean. . . ."

March 5, 1832, St. Jago: "During the walk I was chiefly employed in collecting numberless small beetles & in geologising. King shot some pretty birds & I a most beautiful large lizard. It is a new & pleasant thing for me to be conscious that naturalizing is doing my duty, & that if I neglected that duty I should at same time neglect what has for some years given me so much pleasure."

April 18, 1832, Rio de Janeiro: "If the eye is turned from the world of foliage above, to the ground, it is attracted by the extreme elegance of the leaves of numberless species of ferns & mimosas. Thus it is easy to specify individual objects of admiration; but it is nearly impossible to give an adequate idea of the higher feelings which are excited; wonder, astonishment & sublime devotion, fill & elevate the mind."

July 1, 1832, Rio de Janeiro: "Attended divine service on board the Warspite: the ceremony was imposing; especially the preliminary parts, such as the 'God save the King,' when 650 men took off their hats. Seeing, when amongst foreigners, the strength & power of one's own nation, gives a feeling of exultation which is not felt at home."

November 4, 1832, Buenos Ayres: "Walked into several of the Churches, & admired the brilliancy of the decorations for which the city is celebrated. It is impossible not to respect the fervor which appears to reign during the Catholic service as compared with the Protestant."

January 1, 1834, Port Desire: "Walked to a distant hill; we found at the top an Indian grave. The Indians always bury their dead on the highest hill, or on some headland projecting into the sea."

November 22, 1835, Tahiti: "The Tahitian service was a very interesting spectacle. The Chapel is a large airy framework of wood; it was filled to excess by tidy clean people of all ages & sexes. I was rather disappointed in the apparent degree of attention; but I believe my expectations were raised too high. Anyhow appearance was quite equal to that in a country Church in England."

Adaptation and Creation

Darwin's whole life was governed by his fascination with the orderliness of living nature—the adaptation of organs to the needs of the organism, and the adaptation of organisms to each other and to their inanimate environment. Soon after the voyage, or perhaps even toward the end of it, Darwin began to search for a scientific evolutionary explanation of this order. Before and during the voyage, in agreement with most of his contemporaries, he relied upon the idea of a Designing Providence. In the long run, Darwin's early insistence on one Creator was transformed into an anti-Creationist argument: a Designing Providence might make as many similar species as He pleased; but in a scientific explanation it would be more parsimonious to assume that similar organisms have evolved from one common progenitor. From the *Diary* we can see that his preoccupation with adaptation never became simply an occasion to sing the praises of the Great Artificer, but always remained a fruitful source of questions and of productive scientific wonder.

January 11, 1832, sailing from Tenerife to Cape Verde Islands: "I am quite tired having worked all day at the produce of my net. The number of animals that the net collects is very great & fully explains the manner so many animals of a large size live so far from land. Many of these creatures, so low in the scale of nature, are most exquisite in their forms & rich colours. It creates a feeling of wonder that so much beauty should be apparently created for such little purpose."

September 13, 1834, Chile, on visiting a gold mine: "I was curious to enquire about the load which each mule carries: on a *level road* the regular cargo weighs 416 pounds. . . . Yet to carry this enormous weight, what delicate slim limbs they have; the bulk of muscle seems to bear no proportion to its power. The mule always strikes me as a most surprising animal: that a Hybrid should possess far more reason, memory, obstinacy, powers of digestion & muscular endurance, than either of its parents.—One fancies art has here out-mastered Nature."

December 29, 1834, Cape Tres Montes: "I took much delight in examining the structure of these mountains. The complicated & lofty ranges bore a noble aspect of durability—equally profitless, however, to man & to all other animals."

January 18, 1836, New South Wales: "A little time before this I had been lying on a sunny bank & was reflecting on the strange character of the animals of this country as compared to the rest of the World. An unbeliever in everything beyond his own reason might exclaim, 'Surely two distinct Creators must have been at work; their object, however, has been the same & certainly the end in each case is complete.' Whilst thus thinking, I observed the conical pitfall of a Lion-Ant: a fly fell in & immediately disappeared; then came a large but unwary Ant. His struggles to escape being very violent, the little jets of sand described by Kirby (Vol. I. p. 425) were promptly directed against him. His fate, however, was better than that of the fly's. Without doubt the predaecious Larva belongs to the same genus but to a different species from the European kind.—Now what would the *Dis*-believer say to this? Would any two workmen ever hit on so beautiful, so simple & yet so artificial a contrivance? It cannot be thought so. The one hand has surely worked throughout the universe. A Geologist perhaps would suggest that the periods of Creation have been distinct & remote the one from the other; that the Creator rested in his labor."

Extracts from the B-C-D-E Transmutation Notebooks

Anticipating social Darwinism, Samuel Johnson wrote that everything ought to be persecuted in order that we may know whether it is worthy to live or not. This notion that submission to the ordeal of fire is the best road to virtue has a medieval ring to it. Modern evolutionary theory, from Darwin on, while emphasizing the importance of struggle, also suggests that a new thing, whether a mutant organism or a deviant idea, needs a safe haven for its formative period. For Darwin's early thoughts about evolution, his notebooks served as the private place in which he could try out new ideas without fear of premature exposure to the fires of criticism.

In July 1838 Darwin had finished the first two transmutation notebooks. They included many notes on man, mind, and materialism. He began the third transmutation notebook at the same time as the M notebook, with the intention of separating these two lines of thought more clearly. But the transmutation notebooks continued to serve as repositories for some of his thoughts about man. His thinking about evolution in general and about man's place in nature were too thoroughly intertwined for him to keep them apart.

In this section we have selected some passages from the transmutation notebooks, both to give the flavor of his early evolutionary thinking and to illustrate the way in which these thoughts were suffused with his reflections on man.

B Notebook: Transmutation of Species

This Book was commenced about July, 1837. p. 235 was written in January 1838, perhaps ended in beginning of February.

Zoonomia

Two kinds of generation the coeval kind, all individuals absolutely similar; for instance fruit trees, probably polypi, gemmipares propagation, bisection of Planariae, etc. etc.—

The ordinary kind which is a longer process, the new individual passing through several stages (?typical <of the> or shortened repetition of what the original molecules has done).—This appears | highest office in organization (especially in lower animals, where mind, & therefore relation to other life has not come into play) —see Zoonomia[1] arguments, fails in hybrids where every thing else is perfect: mother apparently only born to breed.— annuals rendered perennial. etc etc.— ((Yet Eunuch, nor /cut/ Stallions, nor nuns are longer lived))

B2

Why is life short, why such high object—generation.—

We *know* world subject to cycle of change, temperature & all circumstances which | influence living beings.—

B3

we see <living things> the young of living beings, become permanently changed or subject to variety, according (to) circumstance,—seeds of plants sown in rich soil, many kinds are produced, though new individuals produced by buds are constant, hence we see generation here seems a means to vary or adaptation. —Again we <believe> /know/ in course of generations even mind & instinct becomes influenced.— | child of savage not civilized man.—birds rendered wild through generations acquire ideas ditto. V. Zoonomia.—

B4

There may be unknown difficulty with *full grown* individual /with fixed organisation/ thus being modified,—therefore generation to adapt & alter the race to *changing* world.—

On other hand, generation destroys the effect of accidental injuries, which if animals lived for ever would be endless | (that is with our persent systems of body & universe.—therefore final cause of life

B5

With this tendency to vary by generation, why are species all constant over whole country. beautiful law of intermarriages <separating> partaking of characters of both parents, & then *infinite* in number

In man it has been | said, there is instinct for opposites to like each other

B6

Aegyptian cats & dogs ibis same as formerly but separate a pair & place them on fresh isl[d] it is very doubtful whether they would remain constant; is it not said that marrying in *deteriorates* a race, that is alters it from some end which is good for man.— |

Let a pair be introduced & increase slowly, from many enemies, so as often to intermarry who will dare say what result

B7

According to this view animals, on separate islands, ought to become different if kept long enough /apart, with slightly dif-fer[ent] circumstances/.—Now Galapagos Tortoises, Mocking birds, Falkland Fox, Chiloe fox.—English & Irish Hare.— |

B8   As we thus believe species vary, in changing climate we ought to find representative species; this we do in South America closely approaching.—but as they inosculate, we must suppose the change is effected at once,—something like a variety produced— (every

B9   grade in that case surely is not | produced?—

<Granting> Species according to Lamarck[2] disappear as collec-tion made perfect.—truer even than in Lamarck's time. Gray's[3] re-mark, best known species (as some common land shells) most difficult to separate. Every character continues to vanish, bones, in-stinct etc etc etc |

B10   Non-fertility of hybridity etc etc

<assuming all> if species (1) may be derived from form (2) etc.,—then (remembering Lyells arguments of transportal) <con-tinent> island near continents might have some species same as

B11   nearest land, which were late arrivals | others old ones, (of which none of same kind had in interval arrived) might have grown al-tered Hence the type would be of the continent though species all different.

In cases as Galapagos and Juan Fernandez. |

B17   As I have before said *isolate* species, <& give even less change> especially with some change probably <change> vary quicker.—

B18   Unknown causes of change. Volcanic isl[d]—Electricity. | Each species changes does it progress.

man gains ideas.

the simplest cannot help becoming more complicated; & if we look to first origin there must be progress.

if we suppose monads are /constantly/ formed, ?would they not

B19   be pretty similar over whole world under | similar climates & as far as world has been uniform, at former epoch. How is this Ehren-berg?[4]

every successive animal is branching upwards different types of organization improving as Owen[5] says simplest coming in & most perfect (& others) occasionally dying out; for instance, secondary

B20   terebratula may | have propagated recent terebratula, but Mega-therium nothing.

We may look at Megatherium, armadillos & sloths as all off-springs of some still older types. some of the branches dying out.—

with this tendency to change (& to multiplication when iso-

B21   lated,) requires deaths of species to keep numbers | /of forms/ equable. ((but is there any reason for supposing number of forms equable: this being due to subdivisions & amount of differences, so forms would be about equally numerous.))

.changes not result of will of animal, but law of adaptation as much as acid & alkali.

Organized beings represent a tree, *irregularly branched* some branches far more branched,—Hence Genera.—As many terminal B22 buds dying, as new one generated. | There is nothing stranger in death of species, than individuals

If we suppose monad definite existence, as we may suppose in this case, their creation being dependent on definite laws; then those which have changed most, /owing to the accident of posi- B23 tions/ must in each state of existence have shortest | life. Hence shortness of life of Mammalia.—

Would there not be a triple branching in the tree of life owing to three elements air, land & water, & the endeavour of each typi- cal class to extend his domain into the other domains & subdivision B24 <six> three more, double arrangement.— | If each main stem of the tree is adapted for these three elements, there will be certainly points of affinity in each branch.

A species as soon as once formed by separation or change in part of country, repugnance to intermarriage <increases>—settles it. |

B25 ?We need not think that fish & penguins really pass into each other.—

The tree of life should perhaps be called the coral of life, base of branches dead; so that passages cannot be seen.—

B26 this again offers | ((no only makes it excessively complicated)) contradiction to constant succession of germs in progress. [Tree Diagram. See page 142 above]

Is it thus fish can be traced right down to simple organization. —birds—not. | [Tree Diagram. See page 142 above]

B27 We-may fancy according to shortness of life of species that in perfection, the bottom of branches deader—so that in mammalian /birds/ it would only appear like circles; & insects amongst articu- lata—but in lower classes perhaps a more linear arrangement.—[6] |

B28 ?How is it there come aberrant species in each genus ((with well characterized parts belonging to each)) approaching another.

<Petrels have divided themselves into many species, so have the awks, there is particular circumstances, to which.> is it an index of the point whence two favourable points of organization com- menced branching.— |

B29 As all the species of some genera have died, have they all one determinate life dependent on genus, the genus upon another, whole class would die out therefore |

B30   In isl[d] neighbouring continent where some species have passed
over, & where other species have "air" of that place, will it be said
B31 those have been then created there:— | Are not all our /British/
shrews diff[erent] species from the continent. Look over Bell[7] and
L. Jenyns.[8] Falkland rabbit may perhaps be instance of domesticated animals having effected, a change which the Fr. naturalists
thought was species. Study Lesson[9]—Voyage of Coquille.— |

B32   Dr Smith[10] say he is certain that when White Men & Hotten-
tots or Negroes cross at C. of Good Hope the children cannot be
made intermediate. the first children partake more of the mother,
the later ones of the father; ((is not this owing to each copulation
producing its effect; as when bitches' puppies are less purely bred
owing to having once born mongrels.)) he has thus seen the black
blood come out from the grandfather (when the mother was nearly
quite white) in the two first children. How is this in West Indies
— ((Humboldt,[11] New Spain.—)) |

B33   Dr. Smith always urges the distinct locality or metropolis of
every species: believes in repugnance in crossing of species in wild
state.—

No doubt /C.D./ wild men do not cross readily, distinctness of
tribes in T. del Fuego.[12] the existence of whiter tribes in centre
of S. America shows this.[13]—Is there a tendency in plants hybrids
to go back?—If so man & plants together would establish law. as
above stated: no one can doubt that less trifling differences are
B34 blended by | intermarriages, then the black & white is so far gone,
that the species (for species they certainly are according to all
common language) will keep to their type: in animals so far removed with instinct in lieu of reason, there would probably be
repugnance & art required to make marriage.—As Dr Smith remarked man and /wild/ animals in this respect are differently
circumstanced.— |

B35   ?Is the shortness of life of *species* in certain orders connected
with gaps in the *series of connection?* ((if starting from same
epoch *certainly*)) The absolute end of certain forms from considering S. America (*independent of external causes*) does appear
very probably:—Mem.: Horse, Llama, etc etc—

If we <suppose> grant similarity of animals in one country owing to springing from one branch, & the monucle has definite life,
then all die at one period, which is not case . . MONUCLE NOT DEF-
INITE LIFE[14] |

I think [Tree Diagram. See page 143 above]

B36   ((Case must be that one generation then should have as many
living as now   To do this & to have many species in same genus
(as is) , REQUIRES extinction.))

Thus between A. & B. immense gap of relation, C & B, the finest gradation, B & D rather greater distinction. Thus genera would B37 be formed,—bearing relation | to ancient types,—with several extinct forms, for if each species /as ancient (1) / is capable of making, 13 recent forms.—Twelve of the contemporarys must have left no offspring at all, so as to keep number of species constant.—

With respect to extinction we can easy see that variety of ostrich Petise may not be well adapted, & thus perish out, or on other B38 hand like Orpheus being favourable, | many might be produced. This requires principle that the permanent varieties produced by <inter> confined breeding & changing circumstances are continued & produce according to the adaptation of such circumstances, & therefore that death of species is a consequence (contrary B39 to what would appear from America) | of non-adaptation of circumstances.—
Vide two pages back Diagram.

The largeness of present genera renders it probable that many contemporary, would have left scarcely any type of their existence in the present world.—or we may suppose only each species in each generation only breeds, *like* individuals in a country not rapidly increasing.— |
B40 If we thus go very far back to look to the source of the Mammalian type of organization; it is extremely improbable that any of <his relations shall likewise> the successors of his relations shall now exist.—In same manner, if we take /a man from/ any large family of 12 brothers & sisters in a state which does not inB41 crease, | it will be chances against any one /of them/ having progeny living ten thousand years hence; because at present day many are relatives, so that by tracing back the <descen> fathers would be reduced to small percentage: therefore the chances are excessively great, against any two of the 12 having progeny after that distant period.— |
B42 Hence if this is true, that the *greater the groups* the *greater the gaps* (or *solutions* of *continuous structure*) /between them./—for instance there would be great gap between birds & mammalia, B43 Still greater between | vertebrate & articulata, still greater between animals & Plants

But yet besides affinities from three elements, from the /infinite/ variation, & all coming from one stock & obeying one law, they may approach,—some birds may approach animals, & some of the vertebrate invertebrate.—Such a few on each side will yet present B44 some anomaly & bearing | stamp of <some> great main type, & the gradation will be sudden.—
Heaven know whether this agrees with Nature: *Cuidado* |

B84   When one sees nipple on man's breast, one does not say some use, but sex not having been determined.—so with useless wings under elytra of beetles.—born from beetles with wings & modified —if simple creation, surely would have been born without them.— |

B93   Man has no *hereditary prejudices* /or instinct/ to conquer or breed together.—Man has no limits to desire, in proportion instinct more, reason less, so will aversion be |

B101   Astronomers might formely have said that God ordered, each planet to move in its particular destiny.—In same manner God orders each animal created with certain form in certain country, but how much more simple, & sublime power let attraction act according to certain law such are inevitable consequences let animal be created, then by the fixed laws of generation, such will be their successors |

B102   Let the powers of transportal be such & so will be the forms of one country to another.—let geological changes go at such a rate, so will be the number & distribution of the species!! |

B118   F. Cuvier[15] says, "But we could only produce domestic individuals & not races, without the occurrence of one of the most general laws of life, the transmission of a fortuitous modification, into a durable form, of a fugitive want into a fundamental propensity, of an accidental habit into an instinct." Ed. N. Phi. J., p. 297, No. 8, Jan.–Apr. 1828.—I take higher grounds & say life is short for this object & others, viz. not too much change. |

B119   In number 6 ? of Ed. n. Phil. Journ. Paper by Crawford[16] on Mission to Ava, account of HAIRY /because ancestors hairy/ man with one hairy child, and of *albino* /DISEASE/ being banished, & given to Portuguese priest.—In first settling a country, people very apt to be split into many isolated races! are there any instances of peculiar people banished by rest?—∴ Most monstrous form has tendency to propagate as well as diseases. |

B142   Parasites of negroes different from European.—Horse & ox have different parasites in different climates.—[17] |

B146   If population of place be constant /say 2000/ and at present day, every ten living souls on average are related to the (200^{dth} year) degree. Then 200 years ago, there were 200 people living who now have successors. Then the chance of 200 people, <might be> being related within 200 years backward might be calculated & this number eliminated say 150 people four hundred years since were progenitors of present people, and so on backwards to one progenitor, who might have continued breeding from eternity backwards.— |

B147   If population was increasing between each lustrum, the number related at the first start must be greater, & this number would vary at each lustrum, & the calculation of chance of the relationship of the progenitors would have different formula for each lustrum.—

We may conclude that there will be a period though long distant, when of the present men (of all races) not more than a few will have successors. At present day in looking at two fine families B148 one will | [have] successors /for/ centuries, the other will become extinct.—Who can analyse causes, dislike to marriage. hereditary disease, affects of contagions & accidents: yet some causes are evident, as for instance one man killing another.—So is it with *varying* races of man: then races may be overlooked mere variations consequent on climate etc.—the whole races act towards each other, and are acted on, just like the two families ((no doubt a different set of causes must act in the two cases)). |

B169   <Angels> (<Races>) Man in *savage* state may be called <species> /races/ in *domesticated* <species> /races/—If all men were dead then monkeys make men.—Men make angels.— |

B207   People often talk of the wonderful event of intellectual man appearing.—the appearance of insects with other senses is more wonderful. its mind more different probably & introduction of man nothing compared to the first thinking being, although hard to B208 draw line— | not so great as between perfect insects & Forms hard to tell whether articulate or intestinal, or even a mite.—A bee /compared with cheese mite/ with its wonderful instincts, <might well say know> The difference is that there is wide gap between man & next animals in mind, more than in structures.

If the skeleton of a negro had been found, what would Anatomists have said.— |

B214   The difference [in] intellect of man & animals not so great as between living thing without thought (plants) & living thing with thoughts (animal).

((∴ My theory very distinct from Lamarcks))

B215   Without *two* species will generate common kind, which is not probable, then monkeys will never produce man, but | both monkeys & man may produce other species, man already has produced marked varieties & may someday produce something else, but not probable owing to mixture of races.—When all mixed & physical changes (?intellectual being acquired alters case) other species or angels produced. |

B216   Has the Creator since the Cambrian formation gone on creating animals with same general structure.—Miserable limited view.—

With respect to how species are, Lamarcks "willing" doctrine absurd (as equally are arguments against it—namely how did otter live before being made otter—why to be sure there were a thousand intermediate forms. |

B227   With belief of <change> transmutation & geographical grouping we are led to endeavour to discover *causes* of change,—the manner of adaptation (wish of parents??), instinct & structure become full of speculation & line of observation.—View of generation being condensation, test of highest organization intelligible.—

B228   May look to first germ, | —led to comprehend true affinities. My theory would give zest to /recent & fossil/ Comparative Anatomy: it would lead to study of instincts, heredity & mind heredity, whole metaphysics.—it would lead to closest examination of hybridity,—& generation, causes of change /in order/ to know what we have come from & to what we tend.—to what circumstances favour crossing & what prevents it this & /direct/ examination of direct passages of <species> structure in species might lead to laws of change, which would then be main object of study, to guide our

B229   speculations | with respect to past and future.

((The grand Question which every naturalist ought to have before him when dissecting a whale or classifying a mite, a fungus or an infusorian is "What are the Laws of Life")) |

B231   Animals whom we have made our slaves we do not like to consider our equals.— ((Do not slave-holders wish to make the black man other kind)) animals with affections, imitation, fear of death, pain, sorrow for the dead—respect |

We have no more reason to expect the father of man kind, than Macrauchenia yet it may be found:—We must not compare /chance of embedment in/ man in present state with what he is as former species. His arts would not then have taken him over whole world.— |

B232   ((The soul by consent of all is superadded, animals not got it, not look forward)) if we choose to let conjecture run wild, then animals our fellow brethren in pain, disease death & suffering /& famine/, our slaves in the most laborious work, our companion in our amusements, they may partake, from our origin in one common ancestor we may be all netted together.— |

C Notebook: Transmutation of Species

c55   Whewell[18] thinks (p. 642) /anniversary/ speech /Feb. 1838/ thinks gradation between man & animals small point in tracing history of man.—granted.—but if all other animals have been so

formed, then man may be a miracle, but induction leads to other
c56 views.— | Till we know uses of organs clearly, we cannot guess causes of change.—hump on back of cow!! etc. etc. |

c61   Whether species may not be made by a little more vigour being given to the chance offspring who have any slight peculiarity of structure. ((hence seals take victorious seals, hence deer victorious deer, hence males armed & pugnacious all orders; cocks all war-like)) |

c62   All the discussion about affinity & how one order first becomes developed & then another— (according as parent types are present) must follow after there is proof of the non creation of animals.— then argument may be,—subterranean lakes, hot springs etc etc inhabited therefore mud wood [would] be inhabited, then how is
c63 this effected by—for instance, fish being excessively abundant | & tempting the Jaguar to use its feet much in swimming, & every developement giving greater vigour to the parent tending to produce effect on offspring—but WHOLE race of that species must take to that particular habitat.—All structures either direct effect of habit, or hereditary /& combined/ effect of habit,—perhaps in process of change.—Are any men born with any peculiarity, or any race of plants.—Lamarck's willing absurd, ∴ not applicable to plants |

c70   Once grant my theory, & the examination of species from distant countries may give thread to conduct to laws of change of organization!

The little turtle, without its parent running to the water, is a good instance of innate instinct, better than child sucking or even duckling & fowls—

When talking of races of men,—black men, black bull finches— from linseed—not solely effects of climate on some antecedent race, perhaps not one now existing. |

c73 Study the wars of organic being.—the fact of guavas having overrun Tahiti, thistle Pampas show how nicely things adapted—Then /aberrant/ varieties will be formed in any kingdom of nature, where scheme not filled up (most false to say no passages; nature is full off them.—Wading birds partially webbed etc etc) —& in round of chances every family will have some aberrant groups,— but as for number five in each group absurd.—The mere fact of
c74 division of lesser & more power (2. typical 3. subtypical) | where power arbitrary, leaves door open for Quinarians to deceive himself.—

Give the case of Apterix split, depress & elevate & enlarge New Zealand, a division of nature of Apterix, many genera & species— The believing that monkey would breed (if mankind destroyed) some intellectual being though not MAN,—is as difficult to understand as Lyells doctrine of slow movements &c &c |

c75   This multiplication of little means & bringing the mind to grapple with great effect produced is a most laborious & painful effort of the mind (although this may appear an absurd saying) & will never be conquered by anyone (if has any kind of prejudices) <without> who just takes up & lays down the subject without long meditation—His best chance is to have [thought] profoundly over the enormous difficulty of reproduction of species & certainly of destruction; then he will choose & firmly believe in his new faith of the lesser of the difficulties. |

c76   Once grant that /species/ [of] one genus may pass into each other,—grant that one instinct to be acquired (if the medullary point in ovum has such organization as to force in one man the developement of a brain capable of producing more glowing imagery or more profound reasoning than other—if this be granted!!) & whole fabric totters & falls.—look abroad study grada-
c77 tion study unity of type study geographical distribution | study relation of fossil with recent, the fabric falls! But man—wonderful man "divino ore versum coelum attentior" is an exception.—He is mammalian,—his origin has not been indefinite—he is not a deity his end /under present form/ will come. (or how dreadfully we are deceived) then he is no exception.—he possesses some of the same general instincts, & <moral> feelings as animals.—they on other hand can reason—but man has reasoning powers in excess,
c78 instead of | definite instincts—this is a replacement in mental machinery, so analogous to what we see in bodily, that it does not stagger me.—What circumstances may have been necessary to have made man! Seclusion want etc & perhaps a train of animals of hundred generations of species to produce contingents proper.— Present monkeys might not,—but probably would,—the world |
c79 now being fit, for such an animal—man, (rude uncivilized man) might not have lived when certain other animals were alive, which have perished.

Let man visit Ourang-outang in domestication, hear expressive whine, see its intelligence when spoken, as if it understood every word said—see its affection to those it knows,—see its passion & rage, sulkiness & every action of despair; ((let him look at savage, roasting his parent, naked, artless, not improving, yet improvable)) & then let him dare to boast of his proud preeminence.— ((Not understanding language of Fuegian puts on par with monkeys)) |

c83   Peculiarities of structure, as six-fingered people are sometimes hereditary,—yet these not adaptations—((they are counteracted by nature by crossing with other varieties)) but <accidental>

changes after birth do not affect progeny. Many dogs in England must have been lopped off & sheeps tails cut yet there is no record of any effect.—New Hollanders have gone on boring their noses, etc & This congenital changes show that grandson is determined when child is.— |

C123   What the Frenchman did for *species* between England & France, I will do with *forms*. —

Mention persecution of early Astronomers,—then add chief good of individual scientific men is to push their science a few years in advance only of their age, (differently from literary men,) must remember that if they *believe* & do not openly avow their belief they do as much to retard as those whose opinion they be-

C124 lieve have endeavoured to advance cause of truth. | It is of the utmost importance to show that habits sometimes go before structure.—the only argument can be a bird practising imperfectly some habit, which the whole rest of other family practise with a peculiar structure, thus <Milvulus forficatus> /Tyrannus sulphureus/ if compelled solely to fish, structure would alter.— |

C154   Animals have voice, so has man. not *saltus*, but *hiatus* animals expression of countenance. They may convey much thus. ((hence if sickness death, unequal life—stimulated by same passions, brought into the world same way)) Man has expression.—animals signals, (rabbit stamping ground) man signals.—animals understand the language. they know the cry of pain as well as we.—

It is our arrogance, to raise on the same shelf to (look at common ancestor. scarcely conceivable in savages) Has not the white man, who has debased /& violated every such instinctive feeling/ his nature by making slave of his fellow Black, often wished to consider him as other animal.—it is the way of mankind & I believe

C155 those who soar above such prejudices yet have | justly exalted nature of man. like to think his origin godlike, at least every nation has done so as yet.—

We know what is the natural arrangement. It is the classification of <arrangement> relationship, latter word meaning descent.— |

C165   habits become important element in classification, because structure has tendency to follow it, or it may be hereditary & strictly point out affinities. conduct of Gould,[19] remark of D'Orbigny[20] point out importance of habits in classification.— |

C166   Thought (or desires more properly) being hereditary it is difficult to imagine it anything but structure of brain hereditary, analogy points out to this.—love of the deity effect of organization, oh you materialist!—Read Barclay[21] on organization!! Avitism in mental structure a disposition & avitism in corporeal structure are facts full of meaning.—Why is thought being a secretion of brain, more wonderful than gravity a property of matter? It is our arrogance, it our admiration of ourselves.— |

C171 Reflect much over my view of particular instinct being memory transmitted without consciousness ((a most possible thing see man walking in sleep)).—an action becomes habitual is probably first stage, & an habitual action implies want of consciousness & will & therefore may be called instinctive.—But why do some actions become hereditary & instinctive & not others.—We even see they must be done often /to be habitual/ or of great importance to cause long memory,—structure is only gained slowly, therefore it can only be those actions which *many* successive generations are impelled to do in same way.—The improvement of reason implies diversity & therefore would banish individual but general ones might yet be transmitted.— |

C172 Memory springing up after long intervals of forgetfulness,—after sleep /strong/ analogies with memory in offspring.—some association in such cases recall the idea—it is scarcely more wonderful that it should be remembered in next generation. ((or simply structure in brain people in fever recollecting things utterly forgotten)) [(N.B. What are those marvellous cases, when you feel sure you have heard conversation before, is strong association recalling up image which had been past—so great an anomaly in structure of brain not probable) put note Sir W. Scott has written about it]CD If we saw a child do some action, which its father had done habitually we should exclaim it was instinct, even if savage take or was given a great coat & this he put on & we afterwards could understand /language better instance/ he had done this without reflection or consciousness of reasoning to tell back from C173 front etc or use of button holes it would | be instinctive.—My view of instinct explains its loss ? if it explains its acquirement.— analogy a bird can swim without being web footed, yet with much practice & led on by circumstance it becomes web footed. now man by effort of memory can remember how to swim after having once learnt, & if that was a regular contingency the brain would become webfooted & there would be no act of memory.—

[There is no corelation between individual objects as Ichneumon & caterpillar though our ignorance may make us think so, but only between laws.]CD |

C174 Many diseases in common between man & animals. Hydrophobia etc cowpox, proof of common origin of man.—different contagious diseases, where habits of people nearly similar. Curious instance of difference in races of men.—

Wax of Ear, bitter perhaps to prevent insects lodging there, now these exquisite adaptations can hardly be accounted for by my method of breeding there must be some corelation, but the whole mechanism is so beautiful. The corelations are not, however, perfect, else one animal would not cause misery to other,—else smell of man would be disagreeable to mosquitoes. |

c175   We never may be able to trace the steps by which the organization of the eye passed from simpler stage to more perfect preserving its relations.—the wonderful power of adaptation given to organization.—This really perhaps greatest difficulty to whole theory.— |

c190   try to trace from simplest reasoning in lower animals many times produced, a general tendency produced, such as man getting habitually into passion becomes habitually passionate.—the key to affections might perhaps thus be found—a person who is habitually kind to children increases general instinctive feeling.— |

c196   Man in his arrogance thinks himself a great work worthy the
c197   interposition of a deity. More humble & I believe truer to | consider him created from animals.

Insects shamming death. most difficult case to imagine how art acquired.—They reason however on this to a degree. Mem Spider only dropping where growth thick.—Shamming death it is but being motionless. How is instinctive dread ((it is exceedingly doubtful whether animals have any fear of death, only of pain)) of death acquired? The S. American dung beetles will each become the father of many species, a few eggs transported to the St. of Magellan.—Change of habits in Van Diemen's Land. |

c198   Study Mr. Blyth's papers on Instinct.[22]—His distinction between reason & instinct very just, but these faculties being viewed as replacing each other it is hiatus & not saltus.—

The greater individuality of mind in man is analogous to greater individuality of bodies of some animals over those of others.—the mind of different animals less divided.—But as man has hereditary tendencies, his mind is still only a divided body.

P. 3 language seems to supply instincts,—& those powers which allow of acquirement of language, hereditary, acquirable.—therefore man's mind not so different from that of brutes

Hard to say, what is instinct in animals /& what reason/ in precisely same way not possible to say what habitual in man & what reasonable—Some actions may be either in same individual. |

c204   The *races* of men differ chiefly in <size> colour, form of head /& features/ (hence intellect?) & what kinds of intellect) quantity & kind of hair forms of legs—hence the father of mankind probably possessed a structure in these points for a less time than other points.— ((female genital organs.—make abstract on this subject from Lawrence,[23] Blumenbach[24] & Prichard.[25])) ((In some monkeys clitoris wonderfully produced))—Now we might expect that animal half way between men & monkey would have differed in hair, colour & form of head /& features/; but likewise in length of extremities, /how are races in this respect/ upper & lower, which I do not know whether it differs in present races, & form of feet.—Negro /or father of negro/ probably was first black at base of nails & on white of eyes.— |

c212   A monkey (Baboon) at Z. Garden upon being beaten behaved very differently from a dog.—more like a man. continued long in a passion & looked out for him to come again very differently from dog. perhaps being in passion chief difference |

c216e . . . It is capable of demonstration that all animals have never at any one time formed chain, since if cretaceous period assumed, then some perished before. carboniferous some perished | before,

c217   then there always have been gaps, & there now must be, ∴ extinction of species bears relation to existence of general etc etc

Discussion useless, until it were fixed what a species means Two savages, two species,—civilized man may exclaim with Christian /we are all/ Brothers in spirit, all children of one father,—yet differences carried a long way.

c218   Lᵈ Jeffrey[26] (Life of Mackintosh, Vol. II, p. 495) —["] in fact in all reasoning, of which human nature is the object, there is really no natural starting place, because there is nothing more elementary than that complex nature itself with which our speculations must end as well as being" &c &c. their centre is everywhere & their circumference nowhere as long as this is so—!! Metaphysics!!! |

c220   Educate all classes, avoid the contamination of castes. improve the women. (double influence) & mankind must improve—

c237e   ∴ Those animals, which only propagate by scission can not alter much?!

Mr. Brown showed me Bauer's[27] drawings of a curious plant where a tube consisting of pistils & stamens united into long organ, moved on being touched, so as to protect itself, one segment of the corolla being (probably) small to allow it to lie on one side.— but in other species, this segment is converted into hood which

c238e possesses power of movement, & not the organ itself | How except by direct adaptation has such a change been effected.—the consciousness of the plant that this part must be protected however it may be effected.— |

c241e   I suspect some valuable analogies might be drawn between habitual actions of plants (when exciting cause is absent) &

c242e   memory of animals.— (surely in plants | movements effects of irritability, though means injection of fluid different from contraction of fibre) —it is most remarkable habitual action in plants, it allows of any degree in lowest animals—habitual action in intestines subject to sympathetic nerves—

The vividness of first <thoughts> memory in childhood or rather their memory. Very remarkable—scenes in themselves accidental—my first thought of sea side— |

c243   Study Bell on Expression[28] & the Zoonomia, for if the former shows that a man grinning is to expose his canine teeth ((this may be made a capital argument. if man does move muscles for uncovering canines)) no doubt a habit gained by formerly being a baboon with great canine teeth.— ((Blend this argument with his having canine teeth at all.—)) This way of viewing the subject important.—Laughing modified barking, smiling modified laughing. Barking to tell other animals in associated kinds of good news, discovery of prey.—arising no doubt from want of assistance. —crying is a puzzler.—Under this point of view expression /of all animals/ becomes very curious—a dog snarling in play.— |

c244   Hensleigh says the love of the deity & thought of him /or eternity/ only difference between the mind of man & animals.—yet how faint in a Fuegian or Australian! Why not gradation.—no greater difficulty for Deity to choose. when perfect enough for future state, that when good enough for Heaven or bad enough for Hell.— (Glimpses bursting on mind & giving rise to the wildest imagination & superstition.—York Minster story of storm of snow after his brothers murder.—good anecdote.[29] |

c257e   In Holme's History of Man at Maer,[30] it is said the Samoyed women, (?North end of the Oural mountains) have black nipples to their breasts.— |

D Notebook: Transmutation of Species

D36   16th Aug. What a magnificent view one can take of the world Astronomical causes, modified by unknown ones. cause changes in geography & changes of climate superadded to change of climate from physical causes,—then superadded changes of form in the organic world, as adaptation, & these changing affect each other, & their bodies by certain laws of harmony keep perfect in these themselves.—instincts alter, reason is formed & the world peopled /with myriads of distinct forms/ from a period short of eternity

to the present time, to the future.—How far grander than idea
D37 from cramped | imagination that God created (warring against
those very laws he established in all <nature> organic nature)
the Rhinoceros of Java & Sumatra that since the time of the Silu-
rian he has made a long succession of vile molluscous animals. How
beneath the dignity of him, who /is supposed to have/ said let
there be light & there was light ((whom it has been declared "he
said let there be light & there was light"—/bad taste/))

D38 With respect to future destinies of mankind, some of species or
varieties are becoming extinct, others though the negro of Africa is
not loosing ground, yet, as the tribes of the interior are pushing
into each other from slave trade & colonization of S. Africa, so
must the tribes become blended & prevent the strong separation
D39 which | otherwise would have taken place otherwise in 10,000 years
negro probably a distinct species—We know how long a Mammal
may go on as one species from Egyptian mummies & from the
existing animals found fossil when Europe must have worn a quite
different figure

19th With respect to the Deluge, it may be worth adding in note
that amongst the Mammalia of Europe the shells of do—shells of
N. America—shells of S. America.—there is no appearance of sudden termination of existence,—nor is there in the Tertiary
<older> geological epochs.— |

D49 . . . Mayo[31] (Philosophy of Living) quotes Whewell[32] as profound because he says length of days adapted to duration of sleep
in man!! whole universe so adapted!!! & not man to Planets.— in-
stance of arrogance!! |

D111   . . . How long will the wretched inhabitants of N. W. Australia,
go on blinking their eyes without extermination, & change of structure.—When will the musquitoes of S. America take an effect—
would perfect impunity from muskitoes bite influence propagation of species.—

Case of association very disagreeable hearing maid servant cleaning door outside, as often as she touched handle, though really
fully aware she was not coming in,—could not help being per-
fectly disturbed referred to Book M. |

D134e   Sept. 28th We ought to be far from wondering of changes in
numbers of species, from small changes in nature of locality. (I do
not doubt every one till he thinks deeply has assumed that increase of animals exactly proportionate to the number that can
live.—) Even the energetic language of <Malthus> Decandoelle
does not convey the warring of the species as inference from Malthus.— (increase of brutes must be prevented solely by positive
checks, excepting that famine may stop desire.—) in nature pro-

duction does not increase, whilst no check prevail, but the positive check of famine & consequently death.[33] |

D135e population in increase at geometrical ratio in FAR SHORTER time than 25 years—yet until the one sentence of Malthus[34] no one clearly perceived the great check amongst men.— (there is spring, like food used for other purposes as wheat for making brandy.—) Even a *few* years plenty, makes population in men increase, & an *ordinary* crops causes a dearth) take Europe on an average every species must have same number killed year with year, by hawks, by colds etc—even one species of hawk decreasing in number must affect instantaneously all the rest.— (The final cause of all this wedging, must be to sort out proper structure, & adapt it to change. —to do that for form, which Malthus shows is the final effect (by means however of volition) of this populousness, or the energy of man) One may say there is a force like a hundred thousand wedges trying force every kind of adapted structure into the gaps in the oeconomy of nature. or rather forming gaps by thrusting out weaker ones.—

D136e Sept 29th Dr Andrew Smith. (Remarks on extraordinary *curiosity* of Monkeys) The Baboon of which anecdotes have been told is Cynocephalus Porcarius.—this monkey did not like a great coat made for it at first, but in two or three days learn its comfort &

D137 though could not put it *on,* yet threw it over | it, & made it meet in front.—Dr. Smith every baboon & monkey, big & little that ever he saw knew women.—he has repeatedly seen them try to pull up petticoats, & if women not afraid, clasp them round waist & look in their faces & make the st. st noise.—The Cercopithecus *chinensis?* (or bonnet face) monkey he has seen do this.—These monkey had no curiosity to pull up trousers of men. Evidently knew <men> women, thinks perhaps by smell,—but monkeys examine sexes of every |

D138 Has repeatedly seen one he kept pull up feathers of tail of Hen, which lived with it,—also of dogs *but did not seem to evince more lewdness for bitch than dog:* monkeys thus examine each other sexes /by taking up tail/—Mem.: Ourang Jenny with Tommy.— Good evidence of knowledge of woman.—

The noise st st which the C. Sphynx makes is also made by the C. porcarius, together with a grunting noise, the former signifies recognition with pleasure as when food is offered, as much as

D139 to | say give me—the other when Dr Smith more distant.—But he thinks other monkeys make st.—noise.

In case of woman instinctive desire may be said more definite than with bitch, for some feeling must urge them to these actions. ((These facts may be turned to ridicule, or may be thought disgusting, but to philosophic naturalist pregnant with interest))

Hyaena, thinks, when pleased cocks his ears, when frightened depresses them.—

England was united to Continent when elephants lived, & when present animals lived—we know the great time necessary to form channel & (& Basses St.) yet no change in English species—time no element in *making* change, only in *fixing* it: only circumstances a contingency of time. |

D140   When we multiply the effects of <earthquakes>, elevating forces in raising continents, & forming mountain-chains, when we estimate the matter removed by the waves of the sea, on beaches, we really measure the rapidity of change of forms, & instincts in the animal kingdom.—It is the unit of our calendar—epochs & creations reduce themselves to the revolutions of one system in the Heavens.—

Is not *puma* same colour as *lion* because inhabitant of *plain* & Jaguar of woods etc like ground birds. |

D158   Hunter[35] shows almost all animals subject to Hermaphroditism, —those organs which perform nearly same function in both sexes, are never double, only modified, those which perform very different, are both present in very shade of perfection.—How comes it nipples though abortive, are so plain in man, & yet no trace of abortive womb, or ovarium,—or testicles in female.—the presence of both testes & ovaries in Hermaphrodite—but not of paenis & clitoris, shows to my mind, that both are present in every animal, but unequally developed.—surely analogy of Molluscs & neuter bee would shew this. (Do any male animals give milk)—But this D159e not distinctly stated by Hunter,—Do testes, & ovaria when | they first appear occupy their *proper* positions,—this would be argument for developement of either.— (Mammae or sheath of Horses penis reduced to extreme degree of abortion).—Insecta.—hermaphrodite. being not only dimidiate, but quarter grown seems to show whole body imbued with possibility of becoming either sex.—

In my theory I must allude to separation of sexes as very great difficulty, then give speculation to show that it is not overwhelming.— |

D162   Theory of sexes (woman makes bud, man puts primordial vivifying principle) one individual secretes two substances, although organs for the double purpose are not distinguished. (yet may be presumed from hybridity of ferns) afterwards they can be seen distinct (in dioecious plants in their abortive sexual organs?): they then become so related to each other as never to be able to impregnate themselves (this never happens in plants) /only in subordinate manner in the plants which have male & female flowers on same stem.—/ so that Molluscous hermaphroditism takes place.

—thus one organ in each becomes obliterated, & sexes as in Vertebrates take place.—∴ every man & woman is hermaphrodite:—
∴ developed instincts of capon & power of assuming male plumage
in females, & female plumage in castrated male.—Men giving
milk— |

D170   There is an analogy between caterpillars with respect to moths,
& monkeys & men,—each man passes through its caterpillar state.
The monkey represents this state.— |

E Notebook: Transmutation of Species

E46   The dog being so much more intellectual than fox, wolf etc etc
—is precisely analogous case to man exceeding monkeys.— |

E47   Having proved mens & brutes bodies on one type: almost superfluous to consider minds.—as difference between mind of a dog &
a porpoise was not thought overwhelming.—yet I will not shirk
difficulty—I have felt some difficulty in conceiving how inhabitant
of Tierra del Fuego is to be converted into civilized man.—ask the
Missionaries about Australian yet slow progress has done so.—
show a savage a dog, & ask him how wolf was so changed. |

E48   When discussing extinction of animals in Europe, the forms
themselves have been basis of argument of change.—now take
greater area of water & snow-line descent.

My theory gives great final cause ((I do not wish to say only
cause, but one great final cause, nothing probably exists for one
cause)) of sexes /in separate animals/: for otherwise there would
be as many species, as individuals, & though we may not trace out
all the ill effects,—we see it is not the order in this perfect world,

E49   either | at the present, or many anterior epochs.—but we can see
if all species, there would not be social animals. hence not social
instincts, which as I hope to show is /probably/ the foundation of
all that is most beautiful in the moral sentiments of the animated
beings—etc ((this is stated too strongly for there would be innumerable species, & hence few only social there could not be one
body of animals, living with certainty on others))

If man is *one* great object, ((Whether he was or not. He is
present a social animal)) for which the world was brought into
present state,—a feat few will dispute, [although, that it was the
sole object, I will dispute, when I hear from the geologist the
history, & from the Astronomer that the moon probably is unin-
habited]CD & if my theory be true then the formation of sexes
rigidly necessary.— |

E50   Without sexual crossing, there would be endless changes, &

hence no feature would be deeply impressed on it, & hence could not be *improvement* /& hence not in higher animals/—it was absolutely necessary that Physical changes should act not on individuals, but on masses of individuals.—so that the changes should be slow & bear relation to the whole changes of country, & not to the

E51 local | changes—this could only be effected by sexes. All the above should follow after discussion of crossing of <species> individuals with respect to representative species, when going North & South |

E58 Three principles will account for all
  (1) Grandchildren like grandfathers
  (2) Tendency to small change. (especially with physical change)
  (3) Great fertility in proportion to support of parents |

E59 Herschel[36] calls the appearance of new species the mystery of mysteries, & has grand passage upon the problem.! Hurrah—"intermediate causes" |

E63 Are the feet of water-dogs at all more webbed than those of other dogs.—if nature had had the picking she would make <them> such a variety far more easily than man,—though *man's practiced* judgment even without time can do much.— (yet one cross, & the permanence of his breed is destroyed) .

When two races of men meet, they act precisely like two species of animals.—they fight, eat each other, bring diseases to each other etc, but then comes the more deadly struggle, namely which have |

E64 the best fitted organization, or instinct (i.e. intellect in man) to gain the day.—In man chiefly intellect, in animals chiefly organization, though Cont. of Africa & West Indies shows organization in Black Race there gives the preponderance, intellect in Australia to the white.—The peculiar skulls of the men on the plains of Bolivia—strictly fossil—/& in Van Diemen's land/—they have

E65 been exterminated on *principles.* strictly applicable to the | universe—The range of man is not unlike that of animals transported by floating ice.—I agree with Mr Lyell, man is not an *intruder*—: the geological history of man is as perfect as the Elephant, if some genus holding same relation as Mastodon to Man were to be discovered.

Man acts on, & is acted on by /the/ organic and inorganic agents of this earth, like every other animal. |

E89 Jan. 6th The rudiment of a *tail* shows man was originally *quadru*<manous> (*ped*) ((Hairy—could move his ears))

The head being six metamorphosed vertebrae, the parents of all vertebrate animals—must have been like some molluscous /bisexual/ animal with a vertebra only & no head—!!

Handwriting is determined by most complicated circumstances, as shown by difficulty in forging. Yet handwriting said to be hereditary, shows well what minute details of structure of [i.e., are] hereditary. |

E108   Wonderful as is the possession of voice by Man, we should remember, that even birds can imitate the sounds surprisingly well.— |

E114   March 12th. It is difficult to believe in the dreadful /but quiet/ war of organic beings, going in the peaceful woods, & smiling fields.—we must recollect the multitude of plants introduced into our gardens (opportunities of escape for foreign birds & insects) which are propagated with very little care.—& which might spread themselves as well as our wild plants, we see how full nature, how finely each holds its place.—When we hear from authors (Ramond Hort. Transact. Vol. I, p. 17 Append[37]) that in the Pyrenees that

E115e   the | Rhododendron ferrugineum begins at 1600 metres precisely & stops at 2600 & yet know that plant can be cultivated with ease near London—what makes the line, as of trees in Beagle Channel —it is not elements!—We cannot believe in such a line. it is other plants.—a broad border of killed trees would form fringe—but there is a contest & a grain of sand turns the balance.— |

E136   It /may/ be said, that wild animals will vary, according to my Malthusian views, within certain limits, but beyond them not,— argue against this—analogy will certainly allow variation as much as /the/ difference between species,—for instance pidgeons—: ((then comes question of genera))

It certainly appears that swallows have decreased in numbers, what cause?? |

E137   Seeing the beautiful seed of a Bull Rush I thought, surely no "fortuitous" growth could have produced these innumerable seeds, yet if a seed were produced with infinitesimal advantage it would have better chance of being propagated & so etc |

E155   I utterly deny the right to argue against my theory because it makes the world far *older* than what geologists think: it would be doing what | others but fifty years since [did] to geologists,—&

E156   what is older—what relation in duration of planet to our lives— Being myself a geologist, I have thus argued to myself, till I can honestly reject such false reasoning |

Notes by Paul H. Barrett

1. *Zoonomia*, p. 487: "This paternal offspring of vegetables, I mean their buds and bulbs, is attended with a very curious circumstance; and that is, that they exactly resemble their parents, as is observable in grafting fruit-trees, and in propa-

gating flower-roots; whereas the seminal offspring of plants, being supplied with nutriment by the mother, is liable to perpetual variation."

2. Lamarck, *op. cit.*: "According as the production of nature are collected and our museums grow richer, we see nearly all the gaps filled up and the lines of demarcation effaced."

3. Gray, John Edward, "Remarks on the Difficulty of Distinguishing Certain Genera of Testaceous Mollusca by Their Shells Alone, and on the Anomalies in Regard to Habitation Observed in Certain Species," *Philosophical Transactions of the Royal Society of London*, 125:301–310, 1835.

4. Ehrenberg, C. G., "On the Origin of Organic Matter from Simple Perceptible Matter, and on Organic Molecules and Atoms; Together with Some Remarks on the Power of Vision of the Human Eye," *Scientific Memoirs, Selected from the Transactions of Foreign Academies of Science and Learned Societies, and from Foreign Journals* (R. Taylor, ed.) , 1:555–583, 1837. In the article Ehrenberg rejects the belief that the smallest visible animal organisms, which he calls the Monad, could arise by spontaneous generation. The word *monad* also appears in the following passage, in Davy, *op. cit.*: "The external world or matter is to us in fact nothing but a heap or cluster of sensations, and in looking back to the memory of our own being, we find one principle which may be called the *monad*, or *self*, constantly present, intimately associated with a particular class of sensations, which we call our own body or organs . . . the monad is always present; we can fix no beginning to its operations, we can fix no limit to them . . . human life may be regarded as a type of infinite and immortal life, and its succession of sleep and dreams as a type of the changes of death and birth to which from its nature it is liable." Darwin read Davy on February 12, 1839. See *N* 62.

5. Probably Richard Owen: in 1836 became first Hunterian Professor of Comparative Anatomy and Physiology at the Royal College of Surgeons; in 1840 authored "Fossil Mammalia," Part I, *Zoology of the Voyage of the Beagle*, Charles Darwin, ed. Following publication of the *Origin of Species*, Owen became one of Darwin's "chief enemies."

6. Darwin drew a vertical line in the margin beside the passage from "it would only appear" to "arrangement."

7. Bell, Thomas, *A History of British Quadrupeds, Including the Cetacea*, Van Voorst, London, 1837.

8. Jenyns, Leonard, *A Manual of British Vertebrate Animals: or, Descriptions of all the Animals . . . Observed in the British Islands, etc.*, J. Smith, Cambridge, 1835, and *A Systematic Catalogue of British Vertebrate Animals*, Cambridge, 1835.

9. Lesson, René-Primevère, *Voyage autour du monde . . . sur . . . la Coquille*, Paris, 1838–1839.

10. Smith, Andrew, author of *Illustrations of the Zoology of South Africa . . . Collected During an Expedition into the Interior of South Africa, in the Years 1834–1836; etc.*, 5 pt., Smith, Elder, London, 1849 [38–49]. Darwin visited Smith at Cape Town, South Africa, in June 1836.

11. Humboldt, Friedrich Heinrich Alexander von, *Political Essay on the Kingdom of New Spain*, transl. from the original French by John Black, 2 vols., London, 1811.

12. Darwin, *Voyage of the Beagle*, 1839, *op. cit.*, p. 236: "The tribes have no government or head, yet each is surrounded by other hostile ones, speaking different dialects; and the cause of their warfare would appear to be the means of subsistence."

13. See Barlow, 1934, *op. cit.*, p. 172, for discussion of the Chascas, a tribe of tall, fair-skinned Indians whose young men chose death rather than betray their countrymen.

14. Kirby, 1835, Vol. 2, p. 2, uses the word *Monoculus* in a discussion of the peculiar taxonomic affinities of barnacles. He says that "Linné" considered barnacles as a single genus—*Lepas*—whereas Lamarck regarded them as a Class, the Cirrhipeda. Thus Lamarck, "by the insertion of the aspirate, . . . made his term, like *Monoculus*, half Greek and half Latin. . . ." Here is a possible clue to the origin of Darwin's interest in barnacles, in the research of which he was to spend eight years. Note also that it was to J. E. Gray, mentioned above (n. 3) , that Darwin attributed the transfer of his interest from anatomy to the taxonomy of bar-

nacles (see Darwin, Charles, *A Monograph on the Fossil Lepadidae, or Pedunculated Cirripedes of Great Britain*, Palaeontological Society, London, 1851, p. vi) :

15. Cuvier, Frédéric, "Essay on the Domestication of Mammiferous Animals, with some Introductory Considerations on the Various States in which we may Study their Actions." *Edinburgh New Philosophical Journal*, 3:303–308, 1827; 4:45–60 & 292–298, 1828.

16. Crawfurd, John, "Account of Mr. Crawfurd's Mission to Ava," *Edinburgh New Philosophical Journal*, 3:359–370, 1827.

17. Kirby, *op. cit.*, Vol. 1, p. 85.

18. Whewell, William, "Address to the Geological Society, Delivered at the Anniversary, on the 16th of February, 1838, by the Rev. William Whewell, M.A., F.R.S., President of the Society," *Proceedings of the Geological Society of London*, 2:624–649, 1838, p. 642: "The gradation in form between man and other animals, a gradation which we all recognise, and which, therefore, need not startle us because it is presented under a new aspect, is but a slight and, as appears to me, unimportant feature, in looking at the great subject of man's origin."

19. Gould, John, *The Birds of Europe*, 5 vols., Taylor, London, 1837.

20. d'Orbigny, Alcide Dessalines, *Voyage dans l'Amérique Méridionale, etc.*, 9 vols., Pitois-Levrault, Paris, 1835–1847.

21. Barclay, John, *An Inquiry into the Opinions, Ancient and Modern, Concerning Life and Organization*, Bell & Bradfute, Edinburgh, 1822.

22. Blyth, Edward, "On the Psychological Distinctions between Man and all other Animals; and on the Consequent Diversity of Human Influence over the Inferior Ranks of Creation, from any Mutual and Reciprocal Influence exercised among the Latter." *Magazine of Natural History, and Journal of Zoology, Botany, Mineralogy, etc.* (2nd Series Charlesworth), London, 1:1–9, 77–85; 131–141, 1837.

23. Lawrence, William, *Lectures on Physiology, Zoology, and the Natural History of Man, Delivered at the Royal College of Surgeons*, Callow, London, 1819.

24. Blumenbach, Johann Friedrich, *A Manual of the Elements of Natural History*, transl. from 10th German ed. by R. T. Gore, Simpkin & Marshall, London, 1825.

25. Prichard, James Cowles, *Researches into the Physical History of Mankind*, 5 vols., 3rd ed., Sherwood, London, 1836–1847.

26. Jeffrey, Lord Francis, in: *Memoirs of the Life of the Right Honourable Sir James Mackintosh*, Robert James Mackintosh, ed., 2 vols., Moxon, London, 1835.

27. Bauer, Ferdinand, in Matthew Flinders, *A Voyage to Terra Australis . . . in the Years 1801, 1802 and 1803*, 2 vols., Nicol, London, 1814, Vol. 2.

28. Bell, Charles, *Essays on the Anatomy of Expression in Painting*, Longmans, London, 1806.

29. Darwin, Charles, *The Descent of Man, and Selection in Relation to Sex*, 2 vols., Murray, London, 1871, Vol. 1, p. 67: "[York Minster] related how, when his brother killed a 'wild man,' storms long raged, much rain and snow fell."

30. Home, Lord Kames Henry, *Sketches of the History of Man, Considerably Enlarged, etc.*, 4 vols., Creech, Edinburgh, 1788.

31. Mayo, 1838, *op. cit.*

32. Whewell, William, *On Astronomy and General Physics, Considered with Reference to Natural Theology: The Bridgewater Treatises on the Power Wisdom and Goodness of God as Manifested in the Creation*, Pickering, London, 1836, p. 39.

33. Malthus, *op. cit.*, Vol. 1, p. 15: "The positive checks to population are extremely various, and include every cause, whether arising from vice or misery, which in any degree contributes to shorten the natural duration of human life. Under this head, therefore, may be enumerated all unwholesome occupations, severe labor, bad nursing of children, great towns, excesses of all kinds, the whole train of common diseases and epidemics, wars, plague, and famine."

34. *Ibid.*, p. 6: "It may safely be pronounced, therefore, that the population, when unchecked, goes on doubling itself every twenty-five years, or increases in a geometric ratio."

35. Hunter, *op. cit.*, Vol. 4, p. 36.

36. Babbage, Charles, *The Ninth Bridgewater Treatise: A Fragment*, New Im-

pression of the Second Edition, Cass, London, 1967 (Original, Murray, London, 1838) , extract of letter from Sir J. Hershel to Lyell, February 20, 1836, p. 226: "You have succeeded,. too, in adding dignity to a subject already grand . . . where it seems impossible to venture without experiencing some degree of that mysterious awe which the sybil appeals to, in the bosom of Aeneas, on entering the confines of the shades—or what the Maid of Avenel suggests to Halbert Glendinning,

'He that on such quest would go, must know nor fear nor failing;
To coward soul or faithless heart the search were unavailing.'

Of course I allude to that mystery of mysteries, the replacement of extinct species by others."

37. Ramond de Carbonnières, Louis-Elizabeth, "On the Vegetation of High Mountains," translated by Richard Anthony Salisbury (read 2nd April, 1811) , *Transactions of the Horticultural Society of London*, 1 (Appendix) :15–23, 1812.

A Biographical Sketch
of an Infant

Darwin published his paper on early child development in 1877, 37 years after he wrote the diary on which it was based. On the surface it seems as though the recent appearance of Taine's article[1] had simply reminded Darwin of his own notes, with little other connection between the two, especially since Darwin's paper was largely about emotional expression, while Taine's was about the development of language. But there is probably a more interesting historical connection between the two papers.

In the *Descent of Man* Darwin had emphasized the role of emotional expression in the development of language, and he had insisted upon the existence of rudimentary forms of communication in other animals as the evolutionary precursors of human language. Professor Max Müller had disagreed, arguing that there could be no language without thought, that animals could neither think nor speak, and that there was an impassible chasm between man and animals.[2]

Taine had read Müller's critique of Darwin and referred to it in his paper. In spite of certain differences in emphasis, Taine and Darwin were agreed on the natural origins of language, and on the development of language through the child's own activity. Darwin, in publishing his paper in *Mind,* was entering an ongoing controversy, one which his own theory of evolution had provoked. He was seizing an opportunity to defend the idea of the natural origins of all psychological functions.[3]

---

1 H. A. Taine, *op. cit.* See page 224.
2 See P. Giles, "Evolution and the Science of Language," in *Darwin and Modern Science,* edited by A. C. Seward (Cambridge: University Press, 1909) ; Darwin summarized and replied to Müller's arguments in the second edition of *Descent* (1874) .
3 See Max Müller, *Lectures on "Mr. Darwin's Philosophy of Language,"* 1863.

A Biographical Sketch of an Infant[1]

M[onsieur] Taine's very interesting account of the mental development of an infant, translated in the last number of *Mind* (p. 252), has led me to look over a diary which I kept thirty-seven years ago with respect to one of my own infants. I had excellent opportunities for close observation, and wrote down at once whatever was observed. My chief object was expression, and my notes were used in my book on this subject; but as I attended to some other points, my observations may possibly possess some little interest in comparison with those by M. Taine, and with others which hereafter no doubt will be made. I feel sure, from what I have seen with my own infants, that the period of development of the several faculties will be found to differ considerably in different infants.

During the first seven days various reflex actions, namely sneezing, hickuping, yawning, stretching, and of course sucking and screaming, were well performed by my infant. On the seventh day, I touched the naked sole of his foot with a bit of paper, and he jerked it away, curling at the same time his toes, like a much older child when tickled. The perfection of these reflex movements shows that the extreme imperfection of the voluntary ones is not due to the state of the muscles or of the coordinating centres, but to that of the seat of the will. At this time, though so early, it seemed clear to me that a warm soft hand applied to his face excited a wish to suck. This must be considered as a reflex or an instinctive action, for it is impossible to believe that experience and association with the touch of his mother's breast could so soon have come into play. During the first fortnight he often started on hearing any sudden sound, and blinked his eyes. The same fact was observed with some of my other infants within the first fortnight. Once, when he was 66 days old, I happened to sneeze, and he started violently, frowned, looked frightened, and cried rather badly: for an hour afterwards he was in a state which would be called nervous in an older person, for every slight noise made him start. A few days before this same date, he first started at an object suddenly seen; but for a long time afterwards sounds made him start and wink his eyes much more frequently than did sight; thus when 114 days old, I shook a pasteboard box with comfits in it near his face and he started, whilst the same box when empty or any other object shaken as near or much nearer to his face produced no effect. We may infer from these several facts that the winking of the eyes, which manifestly serves to protect

---

[1] *Mind: Quarterly Review of Psychology and Philosophy*, 2:285–294, 1877.

Charles Darwin in 1852, with his eldest child, William. *Courtesy of American Museum of Natural History.*

them, had not been acquired through experience. Although so sensitive to sound in a general way, he was not able even when 124 days old easily to recognise whence a sound proceeded, so as to direct his eyes to the source.

With respect to vision,—his eyes were fixed on a candle as early as the 9th day, and up to the 45th day nothing else seemed thus to fix them; but on the 49th day his attention was attracted by a bright-coloured tassel, as was shown by his eyes becoming fixed and the movements of his arms ceasing. It was surprising how slowly he acquired the power of following with his eyes an object if swinging at all rapidly; for he could not do this well when seven and a half months old. At the age of 32 days he perceived his mother's bosom when three or four inches from it, as was shown by the protrusion of his lips and his eyes becoming fixed; but I much doubt whether this had any connection with vision; he certainly had not touched the bosom. Whether he was guided through smell or the sensation of warmth or through association with the position in which he was held, I do not at all know.

The movements of his limbs and body were for a long time vague and purposeless, and usually performed in a jerking manner; but there was one exception to this rule, namely, that from a very early period, certainly long before he was 40 days old, he could move his hands to his own mouth. When 77 days old, he took the sucking bottle (with which he was partly fed) in his right hand, whether he was held on the left or right arm of his nurse, and he would not take it in his left hand until a week later although I tried to make him do so; so that the right hand was a week in advance of the left. Yet this infant afterwards proved to be left-handed, the tendency being no doubt inherited—his grandfather, mother, and a brother having been or being left-handed. When between 80 and 90 days old, he drew all sorts of objects into his mouth, and in two or three weeks' time could do this with some skill; but he often first touched his nose with the object and then dragged it down into his mouth. After grasping my finger and drawing it to his mouth, his own hand prevented him from sucking it; but on the 114th day, after acting in this manner, he slipped his own hand down so that he could get the end of my finger into his mouth. This action was repeated several times, and evidently was not a chance but a rational one. The intentional movements of the hands and arms were thus much in advance of those of the body and legs, though the purposeless movements of the latter were from a very early period usually alternate as in the act of walking. When four months old, he often looked intently at his own hands and other objects close to him, and in doing so the eyes were turned much inwards, so that he often squinted frightfully. In a fortnight after this time (*i.e.* 132 days old) I observed that if an object was brought as near to his face as his own hands were, he tried to seize it, but often failed; and he did not try to do so in regard to more distant objects. I think there can be little doubt that the convergence of his eyes gave him the clue and excited him to move his arms. Although this infant thus began to use his hands at an early period, he showed no special aptitude in this respect, for when he was 2 years and 4 months old, he held pencils, pens, and other objects far less neatly and efficiently than did his sister who was then only 14 months old, and who showed great inherent aptitude in handling anything.

Anger.—It was difficult to decide at how early an age anger was felt; on his eighth day he frowned and wrinkled the skin round his eyes before a crying fit, but this may have been due to pain or distress, and not to anger. When about ten weeks old, he was given some rather cold milk and he kept a slight frown on his forehead all the time that he was sucking, so that he looked like a grown-up person made cross from being compelled to do something which he did not like. When nearly four months old, and perhaps much earlier, there could be no doubt, from the manner in which the blood gushed into his whole face and scalp, that he easily got into a violent passion. A small cause sufficed; thus, when a little over seven months old, he screamed with rage because a lemon slipped away and he could not seize it with his hands. When eleven months old, if a wrong plaything was given him, he would push it away and beat it; I presume that the beating was an instinctive sign of anger, like the snapping of the jaws by a young crocodile just out of the egg, and not that he imagined he could hurt the plaything. When two years and three months old, he became a great adept at throwing books or sticks, etc., at anyone who offended him; and so it was with some of my other sons. On the other hand, I could never see a trace of such aptitude in my infant daughters; and this makes me think that a tendency to throw objects is inherited by boys.

Fear.—This feeling probably is one of the earliest which is experienced by infants, as shown by their starting at any sudden sound when only a few weeks old, followed by crying. Before the present one was 4½ months old I had been accustomed to make close to him many strange and loud noises, which were all taken as excellent jokes, but at this period I one day made a loud snoring noise which I had never done before; he instantly looked grave and then burst out crying. Two or three days afterwards, I made through forgetfulness the same noise with the same result. About the same time (*viz.* on the 137th day) I approached with my back towards him and then stood motionless: he looked very grave and much surprised, and would soon have cried, had I not turned round; then his face instantly relaxed into a smile. It is well known how intensely older children suffer from vague and undefined fears, as from the dark, or in passing an obscure corner in a large hall, etc. I may give as an instance that I took the child in question, when 2¼ years old, to the Zoological Gardens, and he enjoyed looking at all the animals which were like those that he knew, such as deer, antelopes etc., and all the birds, even the ostriches, but was much alarmed at the various larger animals in cages. He often said afterwards that he wished to go again, but not to see "beasts in houses"; and we could in no manner account for this fear. May we not suspect that the vague but very real fears of children, which are quite independent of experience, are inherited effects of real dangers and abject superstitions during ancient savage times? It is quite conformable with what we know of the transmission of formerly well-developed characters, that they should appear at an early period of life, and afterwards disappear.

Pleasurable Sensations.—It may be presumed that infants feel pleasure whilst sucking, and the expression of their swimming eyes seems to show that this is the case. This infant smiled when 45 days, a second infant when 46 days old; and these were true smiles, indicative of pleasure, for their eyes brightened and eyelids slightly closed. The smiles arose chiefly when looking at their mother, and were therefore probably of mental origin; but this infant often smiled then, and for some time afterwards, from some inward pleasurable feeling, for nothing was happening which could have in any way excited or amused him. When 110 days old he was exceedingly amused by a pinafore being thrown over his face and then suddenly withdrawn; and so he was when I suddenly uncovered my own face and approached his. He then uttered a little noise which was an incipient laugh. Here surprise was the chief cause of the amusement, as is the case to a large extent with the wit of grown-up persons. I believe that for three or four weeks before the time when he was amused by a face being suddenly uncovered, he received a little pinch on his nose and cheeks as a good joke. I was at first surprised at humour being appreciated by an infant only a little above three months old, but we should remember how very early puppies and kittens begin to play. When four months old, he showed in an unmistakable manner that he liked to hear the pianoforte played; so that here apparently was the earliest sign of an aesthetic feeling, unless the attraction of bright colours, which was exhibited much earlier, may be so considered.

Affection.—This probably arose very early in life, if we may judge by his smiling at those who had charge of him when under two months old; though I had no distinct evidence of his distinguishing and recognising anyone, until he was nearly four months old. When nearly five months old, he plainly showed his wish to go to his nurse. But he did not spontaneously exhibit affection by overt acts until a little above a year old, namely, by kissing several times his nurse who had been absent for a short time. With respect to the allied feeling of sympathy, this was clearly shown at 6 months and 11 days by his melancholy face, with the corners of his mouth well depressed, when his nurse pretended to cry. Jealousy was plainly exhibited when I fondled a large doll, and when I weighed his infant sister, he being then 15½ months old. Seeing how strong a feeling jealousy is in dogs, it would probably be exhibited by infants at an earlier age than that just specified, if they were tried in·a fitting manner.

Association of Ideas, Reason, etc.—The first action which exhibited, as far as I observed, a kind of practical reasoning, has already been noticed, namely, the slipping his hand down my finger so as to get the end of it into his mouth; and this happened on the 114th day. When four and a half months old, he repeatedly smiled at my image and his own in a mirror, and no doubt mistook them for real objects; but he showed sense in being evidently surprised at my voice coming from behind him. Like all infants he much enjoyed thus looking at himself, and in less than two months perfectly understood that it was an image; for if I made quite silently any odd grimace, he would suddenly turn around to look at me. He was, however, puzzled at the age of seven months, when being out of doors he saw me on the inside of a large plate-glass window, and seemed in doubt whether or not it was an image. Another of my infants, a little girl, when exactly a year old, was not nearly so acute, and seemed quite perplexed at the image of a person in a mirror approaching her from behind. The higher apes which I tried with a small looking-glass behaved differently; they placed their hands behind the glass, and in doing so showed their sense, but far from taking pleasure in looking at themselves they got angry and would look no more.

When five months old, associated ideas independently of any instruction became fixed in his mind; thus as soon as his hat and cloak were put on, he was very cross if he was not immediately taken out of doors. When exactly seven months old, he made the great step of associating his nurse with her name, so that if I called it out he would look round for her. Another infant used to amuse himself by shaking his head laterally: we praised and imitated him, saying "Shake your head"; and when he was seven months old, he would sometimes do so on being told without any other guide. During the next four months the former infant associated many things and actions with words; thus when asked for a kiss he would protrude his lips and keep still,— would shake his head and say in a scolding voice "Ah" to the coal-box or a little spilt water, etc., which he had been taught to consider as dirty. I may add that when a few days under nine months old he associated his own name with his image in the looking-glass, and when called by name would turn towards the glass even when at some distance from it. When a few days over nine months, he learnt spontaneously that a hand or other object causing a shadow to fall on the wall in front of him was to be looked for behind. Whilst under a year old, it was sufficient to repeat two or three times at intervals any short sentence to fix firmly in his mind some associated idea. In the infant described by M. Taine (pp. 254–256) the age at which ideas readily became associated seems to have been considerably later, unless indeed the earlier cases were overlooked. The facility with which associated ideas due to instruction and others spontaneously arising were acquired, seemed to me by far the most strongly marked of all the distinctions between the mind of an infant and that of the cleverest full-grown dog that I have even known. What a contrast does the mind of an infant present to that of the pike, described by Professor Möbius,[2] who during three whole months dashed and stunned himself against a glass partition which separated him from some minnows; and when, after at last learning that he could not attack them with impunity, he was placed in the aquarium with these same minnows, then in a persistent and senseless manner he would not attack them!

Curiosity, as M. Taine remarks, is displayed at an early age by infants, and is highly important in the development of their minds; but I made no special observation on this head. Imitation likewise comes into play. When our infant was only four months old I thought that he tried to imitate sounds; but I may have deceived myself, for I was not thoroughly convinced that he did so until he was ten months old. At the age of 11½ months he could readily imitate all sorts of actions, such as shaking his head and saying "Ah" to any dirty object, or by carefully putting his forefinger in the middle of the palm of his other hand, to the childish rhyme of "Pat it and pat it and mark it with T." It was amusing to behold his pleased expression after successfully performing any such accomplishment.

I do not know whether it is worth mentioning, as showing something about the strength of memory in a young child, that this one when 3 years and 23 days old on being shown an engraving of his grandfather, whom he had not seen for exactly six months, instantly recognised him and mentioned a whole string of events which had occurred whilst visiting him, and which certainly had never been mentioned in the interval.

Moral Sense.—The first sign of moral sense was noticed at the age of nearly 13 months. I said "Doddy (his nickname) won't give poor papa a kiss,—naughty Doddy." These words, without doubt, made him feel slightly uncomfortable; and at last when I had returned to my chair, he protruded his lips as a sign that he was ready to kiss me; and he then shook his hand in an angry manner until I came and received his kiss. Nearly the same little scene recurred in a few days, and the re-

---

[2] *Die Bewegungen der Thiere, etc.*, 1873, p. 11.

conciliation seemed to give him so much satisfaction, that several times afterwards he pretended to be angry and slapped me, and then insisted on giving me a kiss. So that here we have a touch of the dramatic art, which is so strongly pronounced in most young children. About this time it became easy to work on his feelings and make him do whatever was wanted. When 2 years and 3 months old, he gave his last bit of gingerbread to his little sister, and then cried out with high self-approbation "Oh kind Doddy, kind Doddy." Two months later, he became extremely sensitive to ridicule, and was so suspicious that he often thought people who were laughing and talking together were laughing at him. A little later (2 years and 7½ months old) I met him coming out of the dining room with his eyes unnaturally bright, and an odd unnatural or affected manner, so that I went into the room to see who was there, and found that he had been taking pounded sugar, which he had been told not to do. As he had never been in any way punished, his odd manner certainly was not due to fear, and I suppose it was pleasurable excitement struggling with conscience. A fortnight afterwards, I met him coming out of the same room, and he was eyeing his pinafore which he had carefully rolled up; and again his manner was so odd that I determined to see what was within his pinafore, notwithstanding that he said there was nothing and repeatedly commanded me to "go away," and I found it stained with pickle-juice; so that here was carefully planned deceit. As this child was educated solely by working on his good feelings, he soon became as truthful, open, and tender, as anyone could desire.

Unconsciousness, Shyness.—No one can have attended to very young children without being struck at the unabashed manner in which they fixedly stare without blinking their eyes at a new face; an old person can look in this manner only at an animal or inanimate object. This, I believe, is the result of young children not thinking in the least about themselves, and therefore not being in the least shy, though they are sometimes afraid of strangers. I saw the first symptom of shyness in my child when nearly two years and three months old: this was shown towards myself, after an absence of ten days from home, chiefly by his eyes being kept slightly averted from mine; but he soon came and sat on my knee and kissed me, and all trace of shyness disappeared.

Means of Communication.—The noise of crying or rather of squalling, as no tears are shed for a long time, is of course uttered in an instinctive manner, but serves to show that there is suffering. After a time the sound differs according to the cause, such as hunger or pain. This was noticed when this infant was eleven weeks old, and I believe at an earlier age in another infant. Moreover, he appeared soon to learn to begin crying voluntarily, or to wrinkle his face in the manner proper to the occasion, so as to show that he wanted something.

When 46 days old, he first made little noises without any meaning to please himself, and these soon became varied. An incipient laugh was observed on the 113th day, but much earlier in another infant. At this date I thought, as already remarked, that he began to try to imitate sounds, as he certainly did at a considerably later period. When five and a half months old, he uttered an articulate sound "da" but without any meaning attached to it. When a little over a year old, he used gestures to explain his wishes; to give a simple instance, he picked up a bit of paper and giving it to me pointed to the fire, as he had often seen and liked to see paper burnt. At exactly the age of a year, he made the great step of inventing a word for food, namely, *mum,* but what led him to it I did not discover. And now instead of beginning to cry when he was hungry, he used this word in a demonstrative manner or as a verb, implying "Give me food." This word therefore corresponds with *ham* as used by M. Taine's infant at the later age of 14 months. But he also used *mum* as a substantive of wide signification; thus he called sugar *shu-mum,* and a little later after he had learned the word "black," he called liquorice *black-shu-mum,*—black-sugar-food.

I was particularly struck with the fact that when asking for food by the word *mum* he gave to it (I will copy the words written down at the time) "a most strongly marked interrogatory sound at the end." He also gave to "Ah," which he chiefly used at first when recognising any person or his own image in a mirror, an exclamatory sound, such as we employ when surprised. I remark in my notes that the use of these intonations seemed to have arisen instinctively, and I regret that more observations were not made on this subject. I record, however, in my notes that at a rather later period, when between 18 and 21 months old, he modulated his voice in refusing peremptorily to do anything by a defiant whine, so as to express "That I won't"; and again his humph of assent expressed "Yes, to be sure." M. Taine also insists strongly on the highly expressive tones of the sounds made by his infant before she had learnt to speak. The interrogatory sound which my child gave to the word *mum* when asking for food is especially curious; for if anyone will use a single word or a short sentence in this manner, he will find that the musical pitch of his voice rises considerably at the close. I did not then see that this fact bears on the view which I have elsewhere maintained that before man used articulate language, he uttered notes in a true musical scale as does the anthropoid ape Hylobates.

Finally, the wants of an infant are at first made intelligible by instinctive cries, which after a time are modified in part unconsciously, and in part, as I believe, voluntarily as a means of communication,—by the unconscious expression of the features,—by gestures and in a marked manner by different intonations,—lastly by words of a general nature invented by himself, then of a more precise nature imitated from those which he hears; and these latter are acquired at a wonderfully quick rate. An infant understands to a certain extent, and as I believe at a very early period, the meaning or feelings of those who tend him, by the expression of their features. There can hardly be a doubt about this with respect to smiling; and it seemed to me that the infant whose biography I have here given understood a compassionate expression at a little over five months old. When 6 months and 11 days old he certainly showed sympathy with his nurse on her pretending to cry. When pleased after performing some new accomplishment, being then almost a year old, he evidently studied the expression of those around him. It was probably due to differences of expression and not merely of the form of the features that certain faces clearly pleased him much more than others, even at so early an age as a little over six months. Before he was a year old, he understood intonations and gestures, as well as several words and short sentences. He understood one word, namely, his nurse's name, exactly five months before he invented his first word *mum;* and this is what might have been expected, as we know that the lower animals easily learn to understand spoken words.

APPENDIXES

A Biographical Sketch
of Charles Darwin's Father

Robert Waring Darwin, M.D.
[Biographical sketch by W. Phillips, F.L.S.][1]

Of this family I only know the Doctor Waring Darwin F.R.S. who had had an excellent education and who settled at Shrewsbury when a very young man. He was tall thin and slim then and the first thing that occurred to bring him into notice was the illness of a Mrs. Houlston, wife of a bookseller of Wellington who was attended by Dr.   of the Salop Infirmary. She was dangerously ill. They sent for Dr. who was gone to Birmingham on particular business that detained him two or three days. They were consequently obliged to take up with Dr. Darwin, who, on his arrival at Wellington tried to learn from the Apothecary what medicines he had mixed up for her, and for what disorder she had been treated, but the Apothecary would not give him any information, and he was obliged to act for himself independently. The lady died, and the question was which Doctor killed her. It caused a great sensation, and Dr. Darwin published a pamphlet clearly shewing that Dr.   had treated her for a disease she never had, and the result was he was thought best of. Dr.   left Shrewsbury shortly after and went to Birmingham. Darwin had evidently written his

1 Phillips, W., *Shropshire Men*, 4 (Darwin, pp. 13–17), C57, Shrewsbury Public Library.

pamphlet under a hasty disposition, and therefore, afterwards tried to get as many of them in again as he could.

They were become so scarce that I had not heard of them till I was grown up, and never saw but one, and that was at the sale of Mr. David Parkes's library, which I made sure of, having not then been aware that the Doctor was himself a buyer for the purpose of destroying them—Mr. Panting told me he bought it for 16^s/ or 15^s/. Shortly after the sale the Doctor came to the Old Bank, and after transacting his business said to me: "So you were my competition at Mr. Parkes' Sale," and he then told me that he employed Mr. Panting to bid for him, but added "I had no idea it would go to such a price—you made me pay dearer for it than I ever did for one before."I told him I had been looking out for a copy for a long while. "Oh," says he, "you will look a long while now, for I believe that was the only remaining copy out; at least I do not know of another and the flames have had that."

He became corpulent, and I asked him what his weight was in May 1846, he was however then thinner than he had been, but he got weighed on purpose, and his weight May 30th 1846 was 22 Stone 3½ or 311½ lb.

(He completed his 80th year May 30th 1846 and weighed 23 st. 3½) (note on back of letter)

A tombstone in Montford Church commemorates him and Mrs. Darwin in these words:

Sacred

To the Memory

of Robert Waring Darwin, M.D.

who died on the 13th of November 1848, Aged 82 years

Also

In Memory of Susan his Wife

Who died on the 15 of July 1817 aged 52 years

Her remains are deposited within the Chancel of the Church

Plinian Society Minutes Book[1]
Meeting of 27 March 1827
Extracts from page 57

Mr. Browne then read his paper on organization as connected with Life & Mind in which he endeavoured to establish the following propositions.

    I That all matter is organized

    II That it is the gradually increased perfection in the arrangement of the parts constituting organization, which is the [cause?] [source?] of the distinctions perceptible in the various objects of nature & not specific differences.

    III That Life is the abstract of the qualities inherent in these modes of arranging matter.

    IV That *mind* is to be distinguished from life, being neither one of the functions or combination of qualities, by the concatenation of which life is constituted—nor a term indicating a similar idea.

    And V That mind as far as one individual sense, & consciousness are concerned, is material.

---

1 University of Edinburgh Library. Note: Each line of this entry is crossed out in the original. The announcement that this paper would be read, made at the preceding meeting, was stricken from the record in the same manner.

Acknowledgments

This book has been long in the making and we have needed help from many quarters, always generously given. For permission to print the Darwin notebooks we are grateful to the late Sir Charles Galton Darwin. For permission to reprint excerpts from *Charles Darwin's Diary of the Voyage of H.M.S. "Beagle,"* we thank the editor, Lady Nora Barlow, and Cambridge University Press.

For access to various scattered Darwin manuscripts and related documents we are grateful to many libraries and their staffs, among them the American Philosophical Society, the Athenaeum, the British Museum, the John Crerar Library, the Detroit Public Library, the Library of Congress, the New York Public Library, the Royal Botanic Gardens at Kew, the Shrewsbury Public Library, and Shrewsbury School. University libraries in which we have worked with manuscripts and rare books include the University of Colorado, Cornell University, Edinburgh University, the University of Geneva, Harvard University, Keele University, the University of Michigan, Michigan State University, Oxford University, Rutgers University, and Yale University.

Most of our work has, of course, been done in the Cambridge University Library with its magnificent collection of Darwin materials. Without the assistance of Peter Gautrey this book would have been impossible. We are especially indebted to Dr. Sydney Smith of Cambridge University for sharing unstintingly his matchless knowledge of the

ACKNOWLEDGMENTS

Darwin manuscripts and for the many ways in which he has given us hospitality and practical help.

We also remember pleasant and fruitful days at Down House, Darwin's old home, both wandering about the grounds and sitting in his study using the books that are still kept there. We thank Mr. S. Robinson, custodian of Down House, and Dr. Hedley Atkins of the Royal College of Surgeons for permission to work there.

Lady Nora Barlow has been helpful to us in many ways, in giving us a better feeling for Darwin family traditions, in making important material available to us, and in discussions of Darwin's thought. Dr. June Goodfield, Dr. Stephen Toulmin, and Dr. Robert M. Young all encouraged us to pay close attention to the question of materialism in Darwin's early development. Among the many other individuals with whom we have had valuable discussions of the Darwin materials, or who have helped us in searching for them, we would like to mention J. D. Bernal, Henry Bredeck, Edward Carlin, Gilbert Cohen, Ralph Colp, Gavin de Beer, James J. Doele, Harold Fruchtbaum, Emanuel Hackel, Guy Hamilton, Julian Huxley, Mladen Kabalin, Chester Lawson, James Lawson, Ralph Lewis, Clinton Lockert, Edward Manier, James K. Merritt, Everett Mendelsohn, M. J. Rowlands, Natalia Rubailova, Martin Rudwick, Donald A. Sinclair, Robert Stauffer, Bruce Stewart, George Stocking, Hew Strachan, Arthur Thomas, and Walter A. Weiss. Wilma Barrett and Valmai Gruber have been at numerous points true collaborators in our work.

Howard Gruber is indebted to various institutions for leaves of absence and other forms of help: the University of Colorado, the New School for Social Research, and the Institute for Cognitive Studies at Rutgers University, where he has enjoyed the irreplaceable stimulation of his colleagues. Many students have been helpful in various ways, especially Charles Bebber, Virginia Bernier, Joan Colsey, Donald Hovey, and William Walsh. For financial aid in this work he is indebted to the American Philosophical Society, the Fund for the Advancement of Education, and the National Institute of Mental Health. A year in which to reflect upon creative thinking as a growth process was spent in Geneva at the Centre d'Epistémologie Génétique, due to the hospitality of Professor Jean Piaget.

Paul Barrett is indebted for financial assistance to the All-University Research Fund, the Department of Natural Sciences Royalty Fund, and the Office for Research Development, all of Michigan State University, and to the Arts Fund, Inc., of New York City. He is also grateful to Michigan State University for a sabbatical leave. For encouragement and valuable discussions he thanks the entire faculty of the Department of Natural Science, Michigan State University.

Acknowledgments

223

Colleagues who have read and criticized all or large parts of the book include Solomon E. Asch, Colin Beer, Thomas K. Bever, John Ceraso, Dorothy Dinnerstein, Phillip Liss, Charles St. Clair, John Schmerler, Jacques Vonèche, and Robert M. Young. We are grateful for their comments but do not hold them responsible for our shortcomings.

ILLUSTRATIONS

Page 264: Cynopithecus niger. Drawn from life by Mr. Wolf. Illustrations from *The Expression of the Emotions in Man and Animals.*

Page 311: Charles Darwin and his sister Catherine in 1816. From a colored chalk drawing by Sharples. Courtesy of American Museum of Natural History.

Page 314: M Notebook, p. 74. Courtesy of Cambridge University Library.

Page 327: Cat. Illustrations from *The Expression of the Emotions in Man and Animals.*

Page 330: N Notebook, p. 98. Courtesy of Cambridge University Library.

Page 364: Adaptive behavior in plants. From *The Movements and Habits of Climbing Plants* by Charles Darwin (second edition; London: Murray, 1875, p. 139).

Page 368: N Notebook, p. 36. Courtesy of Cambridge University Library.

Page 383: Spanish Fowl, Polish Fowl. Illustrations from *The Variation of Animals and Plants under Domestication,* vol. 1, pp. 226–229.

Page 415: Darwin marginalia. In John Abercrombie, *Inquiries concerning the Intellectual Powers and the Investigation of Truth* (London: Murray, 1838). Courtesy of Cambridge University Library.

Page 466: Charles Darwin in 1852, with his eldest child, William. Courtesy of American Museum of Natural History.

Index